ANIMAL SCIENCE, ISSUES AND RESEARCH

DOMESTIC ANIMALS

Animal Science, Issues and Research

Additional books and e-books in this series can be found on Nova's website under the Series tab.

ANIMAL SCIENCE, ISSUES AND RESEARCH

DOMESTIC ANIMALS

RICHARD L. ALLEN

snova
New York

Library of Congress Cataloging-in-Publication Data

ISBN: 978-1-53616-312-4

Published by Nova Science Publishers, Inc. † New York

CONTENTS

PREFACE*

This book on the history, breeding, management and diseases of domestic animals is a snapshot of American animal husbandry around the year 1850.

* This is an edited, augmented and reformatted version of "Domestic Animals" by Richard L. Allen, originally published by Agricultural Book Publisher, New York, dated 1865.

INTRODUCTION

The object of the following work, on the History, Breeding, Management, Diseases, &c., of Domestic Animals, is to afford the Stock-breeder and Grazier a connected view of the entire subject in which he has so deep an interest. The writer has endeavored to compress within the limited space assumed as necessary to secure a general circulation and perusal, such principles and practice, and give to each that relative prominence, which it becomes the practical man to observe, to realize the greatest amount of value for the labor and capital devoted to his pursuits.

Their history is essential, as it shows their introduction into the United States, their progress during the various stages of their improvement, and the comparative value of the improved and ordinary breeds. A knowledge of the best mode of breeding and management is of still higher importance. The first will enable the breeder to preserve the high character of the animals in his hands, or perhaps still farther to advance them; while proper management and feeding will prevent that deterioration and loss from disease, which frequently subtract so much from his profits.

A larger space has been purposely devoted to the last topics, in preference to the subject of diseases, as prevention is not only less troublesome than cure, but much more economical. Feeding and management, after breeding, are really the important objects in view to the

Stock-breeder and Grazier, for if these be judiciously attended to, disease among the herds will rarely be known.

The subject of animal diseases is complicated and little understood; and to be properly comprehended, requires years close, intelligent study, under every advantage for obtaining the necessary information. Nearly every disorder assumes various shades of difference, and to remove it effectually a corresponding change of treatment is required. How absurd then the idea, that a compilation of formal remedies, administered by an unskilful or inexperienced manager, will be of material service in rescuing his herds or flocks from the ravages of disease. All that can consistently be done, is to give a few simple remedies for the most common and well-known ailments, and leave to nature or a professional farrier, such as are more complex or unusual.

This work (with many subsequent and important additions) constitutes a small part of the "Compend of American Agriculture," the favorable reception of which, though but recently given to the public, has induced the writer to offer this important division of the subject in its present detached form.

New York, November, 1847

Chapter 1

Introductory Remarks—General Principles of Breeding, Nutrition, Management, &c.

The principal domestic animals reared for economical purposes in the United States, are Horned or neat cattle, the Horse, the Mule, Sheep, and Swine. A few Asses are bred, but for no other object than to keep up the supply of jacks for propagating mules. We have also goats, rabbits, and the house domestics, the dog and cat; the two former, only in very limited numbers, but both the latter much beyond our legitimate wants. There have been a few specimens of the Alpaca imported, and an arrangement is now in progress for the introduction of a flock of several hundred, which, if distributed among intelligent and wealthy agriculturists, as proposed, will test their value for increasing our agricultural resources. We shall confine ourselves to some general considerations, connected with the first-mentioned and most important of our domestic animals.

Their number as shown by the agricultural statistics collected in 1839, by order of our General Government, was 15,000,000 neat cattle; 4,335,000 horses and mules, (the number of each not being specified;) 19,311,000 sheep; and 26,300,000 swine. There is much reason to question the entire accuracy of these returns, yet there is doubtless an approximation to the truth. Sheep have greatly increased since that period, and would probably number, the present year, (1848,) not less than 30,000,000; and if our own manufactures continue to thrive, and we should moreover become wool exporters, of which there is now a reasonable prospect, an accurate return for 1850, will undoubtedly give us not less than 33,000,000 for the entire Union. There has been a great increase in the value of the other animals enumerated, but not in a ratio corresponding with that of sheep. This is not only manifest in their augmented numbers, but in the gradual and steady improvement of the species.

It may be safely predicted, that this improvement will not only be sustained, but largely increased; for there are some intelligent and spirited breeders to be found in every section of the country, whose liberal exertions and successful examples are doing much for this object. Wherever intelligence and sound judgment are to be found, it will be impossible long to resist the effects of a comparison between animals, which, on an equal quantity of the same food, with the same attention and in the same time, will return 50, 20, or even 10 per cent more in their intrinsic value or marketable product, than the ordinary class. This improvement has been, relatively, most conspicuous in the Western and Southern states; not that the present average of excellence in their animals surpasses, or even reaches that of the North and East; but the latter have long been pursuing this object, with more or less energy, and they have for many years had large numbers of excellent specimens of each variety; while with few exceptions, if we exclude the blood-horse or racing nag, the former have, till recently, paid comparatively little attention to the improvement of their domestic animals. The spirit for improvement through extensive sections, is now awakened, and the older settled portions of the country may hereafter expect competitors, whose success will be fully commensurate with their own. Before going into the management of the different varieties, we will give some general principles and remarks applicable to the treatment of all.

The purpose for which animals are required, should be first determined, before selecting such as may be necessary either for breeding or use. Throughout the Northeastern states, cows for the dairy, oxen for the yoke, and both for the butcher, are wanted. In much of the West and South, beef alone is the principal object; while the dairy is neglected, and the work of the ox is seldom relied on, except for occasional drudgery.

Sheep may be wanted almost exclusively for the fleece, or for the fleece and heavy mutton, or in the neighborhood of markets, for large early lambs. The pastures and winter food, climate, and other conditions, present additional circumstances, which should be well considered before determining on the particular breed, either of cattle or sheep, that will best promote the interest of the farmer.

The kind of work for which the horse may be wanted, whether as a roadster, for the saddle, as a heavy team horse, or the horse of all work, must be first decided, before selecting the form or character of the animal.

The range of pig excellence is more circumscribed, as it is only necessary to breed such as will yield the greatest amount of valuable carcass, within the shortest time, and with the least expense.

PRINCIPLES OF BREEDING

All breeding is founded on the principle, that *like begets like*. This is, however, liable to some exceptions, and is much more generally true when *breeding down* than when *breeding up*. If two animals (which can never be exactly similar in all respects) are requisite to the perpetuation of the species, it necessarily results, that the progeny must differ in a more or less degree from each parent. With wild animals, and such of the domestic as are allowed to propagate without the interference of art, and whose habits, treatment, and food are nearly similar to their natural condition, the change through successive generations is scarcely perceptible. It is only when we attempt to improve their good qualities, that it is essential carefully to determine, and rigidly to apply, what are adopted as the present scientific principles of breeding. We cannot believe that we have penetrated beyond the mere threshold of this art. Unless, then, we launch into experiments, which are necessarily attended with uncertainty, our duty will be, to take for our guide the most successful practice of modern times, until further discoveries enable us to modify or add to such as are already known and adopted. We may assume, then, as the present rules for this art,

1st. That the animals selected for breed, should unite in themselves all the good qualities we wish to perpetuate in the offspring.

2d. These qualities, technically called *points*, should be inbred in the animals as far as practicable, by a long line of descent from parents similarly constituted. The necessity for this rule is evident from the fact, that in mixing different species, and especially mongrels, with a long-established breed, the latter will most strongly stamp the issue with its own peculiarities. This is forcibly illustrated in the case of the Devon cattle, an ancient race, whose color, form, and characteristics are strikingly perpetuated, sometimes to the sixth oreven a later generation. So far is this principle carried by many experienced breeders, that they will use an animal of indifferent external appearance, but of approved descent, (*blood*,) in preference to a decidedly superior one, whose pedigree is imperfect.

3d. All the conditions of soil, situation, climate, treatment, and food, should be favorable to the object sought.

4th. As a general rule, the female should be relatively larger than the male. This gives ample room for the perfect development of the fœtus, easy parturition, and a large supply of milk for the offspring, at a period in its existence, when food has a greater influence in perfecting character and form, than at any subsequent time.

5th. Exceptions to this rule may be made, when greater size is required than can be obtained from the female, and especially when more vigor and hardiness of constitution are desirable. For this purpose, strong masculine development in the sire is proper, and if otherwise unattainable, something of coarseness may be admitted, as this may be afterwards corrected, and nothing will atone for want of constitution and strength.

6th. Pairing should be with a strict reference to correcting the imperfections of one animal, by a corresponding excellence in the other.

7th. *Breeding in-and-in*, or propagating from animals nearly allied, may be tolerated under certain circumstances, though seldom; and only in extreme cases between those of the same generation, as brother and sister.

When the animal possesses much stamina and peculiar merit, which it is desired to perpetuate in the breed, it may be done either in the ascending or descending line, as in breeding the son to the parent, or the parent to his own progeny. This has been practised with decided advantage, and in some cases has even been continued successively, as low as the sixth generation.

8th. It is always better to avoid close relationship, by the selection of equally meritorious stock-getters of the same breed, from other sources.

9th. Wholesome, nutritious food, at all times sufficient to keep the animals steadily advancing, should be provided, but they must never be allowed to get fat. Of the two evils, starving is preferable to surfeit. Careful treatment, and the absence of disease, must be always fully considered.

10th. Animals should never be allowed to breed either too early or too late in life. These periods cannot be arbitrarily laid down, but must depend on their time of maturity, the longevity of the breed, and the stamina of the individual.

11th. No violent cross, or mixing of distinct breeds, should ever be admitted for the purposes of perpetuation, as of cattle of diverse sizes; horses of unlike characters; the Merino and the long-wools, or even the long, or short, and the middle-wools. For carcass and constitution, these crosses are unexceptionable; and it is a practice very common in this country, and judicious enough where the whole produce is early destined for the shambles. But when the progeny are designed for breeders, the practice should be branded with unqualified reprehension.

General Form and Characteristics

Within certain limits, these may be reduced to a common standard. All animals should have a good head, well set up; a clean fine muzzle, and a blight, clear and full, yet perfectly placid eye. With the exception of the dog and cat, whose original nature is ferocity, and whose whole life, unless diverted from their natural instincts, is plunder and prey; and the jockey racehorse, which is required to take the purse, at any hazard of life or limb to the groom; a mild, quiet eye is indispensable to the profitable use of the domestic brute. The neck should be well formed, not too long, tapering to its junction with the head, and gradually enlarging to a firm, well-expanded attachment to the back, shoulders, and breast. The back or chine should be short, straight, and broad; the ribs springing out from the backbone nearly at right angles, giving a rounded appearance to the carcass, and reaching well behind to a close proximity to the hip; tail well set on, and full at its junction with the body, yet gradually tapering to fineness; thighs, fore-arms, and crops well developed; projecting breast or brisket; the fore-legs straight, and hind ones properly bent, strong and full where attached to the carcass, but small and tapering below; good and sound joints; dense, strong bones, but not large; plenty of fine muscle in the right places; and hair or wool, fine and soft. The chest in all animals should be full, for it will be invariably found, that only such will do the most work, or fatten easiest on the least food.

The Lungs

From the last-mentioned principle, founded on long experience and observation, Cline inferred, and he has laid it down as an incontrovertible position, that the lungs should always be large; and Youatt expresses the same opinion. This is undoubtedly correct as to working beasts, the horse and the ox, which require full and free respiration, to enable them to sustain great muscular efforts. But later physiologists have assumed, perhaps from closer and more accurate observations, that the fattening propensity is in the ratio of the smallness of the lungs. Earl Spencer has observed, that this is fully shown in the pig, the sheep, the ox, and the horse, whose aptitude to fatten and smallness of lungs, are in the order enumerated.

This position is further illustrated by the different breeds of the same classes of animals. The Leicester sheep have smaller lungs than the South Down; and it has been found, that a number of the former, on a given quantity of food, and in the same time, reached 28 lbs. a quarter, while the South Downs with a greater consumption of food, attained in the same period, only 18 lbs. The Chinese pigs have much smaller lungs than the Irish, and the former will fatten to a given weight, on a much less quantity of food than the latter. (*Playfair.*) The principle would seem to be corroborated by the fact, that animals generally fatten faster in proportion to the quantity of food they consume, as they advance towards a certain stage of maturity; during all which time, the secretion of internal fat is gradually compressing the size, by reducing the room for the action of the lungs. Hence, the advantage of carrying the fattening beast to an advanced point, by which not only the quality of carcass is improved, but the quantity is relatively greater for the amount of food consumed. These views are intimately connected, and fully correspond, with the principles of

Respiration in Animals

From careful experiments, it has been found that all animals daily consume a much larger quantity of food than the aggregate of what may have been retained in the system, added to what has been expelled in the fœces and urine, and what has escaped by perspiration. Boussingault, who combines the characteristics of an ingenious chemist, a vigilant observer, and a practical agriculturist, made an experiment with a "milch-cow and a full-grown horse, which were placed in stalls so contrived that the droppings and the urine could be collected without loss. Before being made the subjects of experiment, the animals were ballasted or fed for a month with the same ration that was furnished to them, during the three days and three nights which they passed in the experimental stalls. During the month, the weight of the animals did not vary sensibly, a circumstance which happily enables us to assume that neither did the weight vary during the seventy-two hours when they were under especial observation.

The cow was foddered with after-math, hay, and potatoes; the horse with the same hay and oats. The quantities of forage were accurately weighed, and their precise degree of moistness and their composition were determined from average samples. The water drunk was measured, its saline and earthy constituents having been previously ascertained. The excrementitious matters passed, were of course collected with the greatest care; the excrements, the urine, and the milk were weighed, and the constitution of the whole estimated from elementary analyses of average specimens of each. The results of the two experiments are given in the table on the next page.

The oxygen and hydrogen that are not accounted for in the sum of the products have not disappeared in the precise proportions requisite to form water; the excess of hydrogen amounts to as many as from 13 to 15 dwts. It is probable that this hydrogen of the food became changed into water by combining during respiration with the oxygen of the air."

Food consumed by the horse in 24 hours

Forage.	**Weight in the wet state**	**Weight in the dry state**		**Elementary Matter in the Food.**													
				Carbon.		Hydrogen.			Oxygen.			Azote.			Salts and Earths.		
	lb.	lb.	oz.	lb.	oz.	lb.	oz.	dwt.	lb.	oz	dwt	lb.	oz	dwt	lb.	oz	dwt
Hay	20	17	4	7	11	0	10	7	6	8	8	0	3	2	1	6	14
Oats	6	5	2	2	7	0	3	18	1	10	14	0	1	7	0	2	10
Water	43	...		...		...			...			...			0	0	8
Total	69	22	6	10	6	1	2	5	8	7	2	0	4	9	1	9	12

Products voided by the horse in 24 hours

Products.	**Weight in the wet state**			**Weight in the dry state**			**Elementary Matter in the Food.**														
							Carbon.			Hydrogen.			Oxygen.			Azote.			Salts and Earths.		
	lb.	oz.	dwt	lb.	oz	dwt	lb.	oz.	dwt	lb.	oz	dwt	lb.	oz	dwt	lb.	oz	dwt	lb.	o	dwt
Urine	3	6	15	9	9	14	0	3	10	0	0	7	0	1	2	0	1	4	0	3	10
Excre-ments	38	2	2	9	5	6	3	7	17	0	5	15	3	6	14	0	2	10	1	6	10
Total	71	8	17	10	3	0	3	11	7	0	6	2	3	7	16	0	3	14	1	10	0
Total matter of the food	69	0	0	22	6	0	10	6	0	1	2	5	8	7	2	0	4	9	1	9	12
Difference	27	3	3	12	3	0	6	6	13	0	8	3	4	11	6	0	0	15	0	0	12

Water consumed by the horse in 24 hours

Water voided by the horse in 24 hours

	lbs.	oz.		lbs.	oz.
With the hay	2	3	With the urine	2	6
With the oats	0	14	With the excrements	23	8
Taken as drink	35	3			
Total consumed	38	4	Total voided	25	14
			Water consumed	38	4
Water exhaled by pulmonary and cutaneous transpiration				12	6

Food consumed by the cow in 24 hours

Fodder	Weight in the wet state			Weight in the dry state			Elementary Matter in the Food.														
							Carbon.			Hydrogen.			Oxygen.			Azote.			Salts and Earths.		
	lb.	oz	d wt	lb.	oz	d wt	lb	oz	d wt	lb.	oz	d wt	lb.	oz	d wt	lb.	oz	d wt	lb	oz	dwt
Pota-toes	40	2	5	11	2	1	4	11	2	0	7	15	4	10	17	0	1	12	0	6	13
After-math hay	20	1	2	16	11	0	7	11	11	0	11	7	5	10	17	0	4	17	1	8	6
Water	160	0	0	...			...			...			...			...			0	1	12
Total	220	3	7	28		1	12	10	13	1	7		10	9	14	0	6	9	2	4	11

Products voided by the cow in 24 hours

Products.	Weight in the wet state			Weight in the dry state			Elementary Matter in the Food.														
							Carbon			Hydrogen			Oxygen			Azote			Salts and Earths		
	lb.	oz	dwt	lb	oz	dwt	lb	oz	dwt	lb	oz	dwt	lb	oz	dwt	lb	oz	dwt	lb	oz	dwt
Excre-ments	76	1	9	10	8	12	4	7	0	0	6	13	4	0	9	0	2	19	1	3	8
Urine	21	11	12	2	6	17	0	8	7	0	0	16	0	8	3	0	1	3	1	0	6
Milk	22	10	10	3	1	0	1	8	3	0	3	3	0	10	6	0	1	9	0	1	16
Total	120	11	11	16	4	9	6	11	10	0	10	12	5	6	18	9	5	11	2	5	10
"matter of food	220	3	7	28	1	1	12	10	13	1	7	2	10	9	14	0	6	9	2	4	11
Difference	99	3	16	11	8	12	5	11	3	0	8	10	5	2	16	0	0	18	0	0	19

We here perceive a large loss of water, carbon, hydrogen, &c. Nearly all this loss of carbon and hydrogen escaped by respiration, while most of the water, oxygen, nitrogen, and salts, passed off in perspiration. In further illustration of the subject of respiration, Liebig says, "from the accurate determination of the quantity of carbon daily taken into the system in the food, as well as of that proportion of it which passes out of the body in the fœces and urine, *unburned*, that is, in some form uncombined with oxygen, it appears that an adult taking moderate exercise, consumes 13.9 oz. of carbon daily." The foregoing are facts in the animal economy, capable of vast practical bearing in the management of our domestic animals. But before following out these principles to their application, let us briefly examine.

Water consumed by the cow in 24 hours **Water voided by the cow in 24 hours**

	lbs.	oz.		lbs.	oz.
With the potatoes	23	12	With the excrements	53	10
With the hay	2	9	With the urine	15	14
Taken as drink	132	0	With the milk	16	3
Total consumed	158	5	Total voided	85	11
			Water consumed	158	5
Water passed off by pulmonary and cutaneous transpiration				79	2

The Effects of Respiration

We have seen from the experiment of Boussingault, that there is a loss of 6 lbs. 6 oz. of carbon, and 8 oz. 3 dwt. of hydrogen in the food of the horse, and something less in that of the cow, every 24 hours, which has not been left in the system, nor has it escaped by the evacuations. What has become of so large an amount of solid matter? It has escaped through the lungs and been converted into air. The carbon and hydrogen of the food have undergone those various transformations which are peculiar to the animal economy, digestion, assimilation, &c., which it is not necessary, nor will our limits permit us here to explain; and they appear at last in the veinous blood, which in the course of its circulation is brought into the cells of the lungs. The air inhaled is sent through every part of their innumerable meshes, and is there separated from the blood, only by the delicate tissues or membranes which enclose it. A portion of the carbon and hydrogen escapes from the blood into the air-cells, and at the instant of their contact with the air, they affect a chemical union with its oxygen, forming carbonic acid and the vapor of water, which is then expired, and a fresh supply of oxygen is inhaled. This operation is again repeated, through every successive moment of animal existence.

Besides other purposes which it is probably designed to subserve, but which have hitherto eluded the keenest research of chemical physiology, one obvious result of respiration is, the elevation of the temperature of the animal system. By the ever-operating laws of nature, this chemical union of two bodies in the formation of a third, disengages latent heat, which taking place in contact with the blood, is by it diffused throughout the whole frame. The effect is precisely analogous to the combustion of fuel, oils, &c., in the open air.

Perspiration

Is the counteracting agent which modifies this result, and prevents the injurious effects, which, under exposure to great external heat, would ensure certain destruction. And this too, it will have been seen, is provided at the expense of the animal food. When from excessive heat, caused by violent exercise or otherwise, by which respiration is accelerated and the animal temperature becomes elevated, the papillæ of the skin pour the limpid fluid through their innumerable ducts, which in its conversion into vapor, seize upon the animal heat and remove it from the system, producing that delicious coolness so grateful to the laboring man and beast in a sultry summer's day. These two opposing principles, like the antagonistic operations of the regulator in mechanics, keep up a perfect balance in the vital machine, and enable that entire division of the animal creation, distinguished as warm-blooded, including man and the brute, all the feathered tribes, the whale, the seal, the walrus, &c., to maintain an equilibrium of temperature, whether under the equator or the poles; on the peaks of Chimborazo, the burning sands of Zahara, or plunged in the depths of the Arctic Ocean.

The connection between the size of the lungs, and the aptitude of animals to fatten, will be more apparent from the fact, that the carbon and hydrogen which are abstracted, constitute two of the only three elements of fat. The larger size, the fuller play, and the greater activity of the lungs, by exhausting more of the materials of fat, must necessarily diminish its formation in the animal system; unless it can be shown, which has never yet been done, that the removal of a portion of the fat-forming principles accelerates the assimilation of the remainder.

The Food Which Supplies Respiration

This, in the herbivorous animals, after they are deprived of the milk, which furnishes it in abundance, is the starch, gum, sugar, vegetable fats, and oils that exist in the vegetables, grain, and roots which they consume; and in certain cases where there is a deficiency of other food, it is sparingly furnished in woody and cellular fibre. All these substances constitute the principal part of dry vegetable food, and are made up of three elements, which in starch, gum, cane-sugar, and cellular fibre, exist in precisely the same proportions, viz: 44 per cent of carbon, 6.2 of hydrogen, and 40.8 of oxygen.

Grape sugar, woody fibre, and vegetable and animal fats and oils are made up of the same elements, but in different proportions, the last containing much more carbon and hydrogen than those above specified. In the fattening animals, it is supposed the vegetable fats and oils are immediately transferred to the fat cells, undergoing only such slight modification as perfectly adapts them to the animal economy, while respiration is supplied by the other enumerated vegetable matters. If these last are taken into the stomach beyond the necessary demand for its object, they too are converted by the animal functions into fat, and are stored up in the system for future use. But if the supply of the latter is insufficient for respiration, it first appropriates the vegetable fat contained in the food; if this is deficient, it draws on the accumulated stores of animal fat already secreted in the system, and when these too are exhausted, it seizes upon what is contained in the tissues and muscle. When the animal commences drawing upon its own resources for the support of its vital functions, deterioration begins; and if long continued, great emaciation succeeds, which is soon followed by starvation and death.

The carnivorous animals are furnished with their respiratory excretions, from the animal fat and fibre which exist in their food, and which the herbivoræ had previously abstracted from the vegetable creation.

Circumstances Which Augment Respiration

These are, exercise, cold, and an abundant supply of food. Exercise, besides exhausting the materials of fat, produces a waste of fibre and tissue, the muscular and nitrogenized parts of the animal system; and it is obvious from the foregoing principles, that cold requires a corresponding demand for carbon and hydrogen to keep up the vital warmth. The consumption of food to the fullest extent required for invigorating the frame, creates a desire for activity, and it insensibly induces full respiration. The well-fed, active man, unconsciously draws a full, strong breath; while the abstemious and the feeble, unwittingly use it daintily, as if it were a choice com modity not to be lavishly expended. If the first be observed when sleep has effectually arrested volition, the expanded chest will be seen, heaving with the long-drawn, sonorous breath; while that of the latter will exhibit the gentle repose of the infant on its mother's breast.

The difference between the food of the inhabitants of the polar and equatorial regions, is strikingly illustrative of the demands both for breathing and perspiration. The latter are almost destitute of clothing, and subsist on their light, juicy, tropical fruits, which contain scarcely 12 per cent of carbon, yet furnish all the elements for abundant perspiration; while the former are imbedded in furs, and devour gallons of train oil or its equivalent of fat, which contains nearly 80 per cent of carbon, that is burnt up in respiration to maintain a necessary warmth.

The bear retires to his den in the beginning of winter, loaded with fat, which he has accumulated from the rich, oily mast abounding in the woods in autumn. There he lies for months, snugly coiled and perfectly dormant; the thickness of his shaggy coat, his dry bed of leaves, and well-protected den, effectually guarding him from cold, which in addition to his want of exercise, draw slightly upon respiration to keep up the vital heat. When the stores of carbon and hydrogen contained in the fat are expended, his hunger and cold compel him to leave his winter-quarters, again to wander in pursuit of food.

Many of the swallow tribes, in like manner, hybernate in large hollow trees, and for months eke out a torpid, scarcely perceptible existence, independent of food. Activity and full respiration, on the return of spring, demand a support, which is furnished in the myriads of flies they daily consume. The toad and frog have repeatedly been found in a torpid state, imbedded in limestones, sandstones, and the breccias, where they were probably imprisoned for thousands of years without a morsel of food; yet when exposed to the warmth of the vital air and the stimulus of its oxygen, they have manifested all the activity of their species. This they are enabled to sustain only by an enormous consumption of insects.

Dr. Playfair states, that in an experiment made by Lord Ducie, 100 sheep were placed in a shed, and ate 20 pounds of Swedes turnips each per day; another 100 were placed in the open air, and ate 25 pounds per day; yet the former, which had one-fifth less food, weighed, after a few weeks, three pounds more per head than the latter. He then fed five sheep in the open air, between the 21st November and 1st December. They consumed 90 pounds of food per day, the temperature being at 44°, and at the end of this time, they weighed two pounds less than when first exposed. Five sheep were then placed under a shed, and allowed to run about in a temperature of 49°. At first they consumed 82 pounds per day; then 70 pounds, and at the end of the time they had gained 23 pounds. Again, five sheep were placed under a shed as before, and not allowed to take any exercise.

They ate at first, 64 pounds of food per day, then 58 pounds, and increased in weight 30 pounds. Lastly, five sheep were kept quiet and covered, and in the dark. They ate 35 pounds per day, and increased eight pounds.

Mr. Childers states, that 80 Leicester sheep in the open field, consumed 50 baskets of cut turnips per day, besides oil-cake. On putting them in a shed, they were immediately able to consume only 30 baskets, and soon after but 25, being only half the quantity required before, and yet they fattened as rapidly as when eating the largest quantity. The minimum of food, then, required for the support of animals, is attained when closely confined in a warm, dark shelter; and the maximum, when running at large, exposed to all weathers.

THE FEEDING OF ANIMALS

This should be regulated by a variety of considerations. The young which may be destined for maturity, should be supplied with milk from the dam until weaning-time. No food can be substituted for the well-filled udder of the parent, which is so safe, healthful, and nutritious. If from any cause there is deficiency or total privation, it must be made up by that kind of food, meal-gruel, &c., which, in its composition, approaches nearest in quality to the milk. At a more advanced age, or the time for weaning, grass, hay, roots, or grain, may be substituted, in quantities sufficient to maintain a steady but not a forced growth. Stuffing can only be tolerated in animals which are speedily destined for the slaughter. Alternately improving and falling back, is injurious to all stock. An animal should never be fat but once. Especially is high feeding bad for breeding animals. Much as starving is to be deprecated, the prejudicial effects of repletion are still greater. The calf or lamb intended for the butcher, may be pushed forward with all possible rapidity. Horses or colts should never exceed a good working or breeding condition.

Purposes Fulfilled by Different Kinds of Food

The objects designed to be answered by food, are to a certain extent the same. All food is intended to meet the demands of respiration and nutrition, and fattening to a greater or less degree. But some are better suited to one object than others, and it is for the intelligent farmer to select such as will most effectually accomplish his particular purposes.

The very young animal requires large quantities of the phosphate of lime for the formation of bone; and this is yielded in the milk in larger proportions than from any other food. The growing animal wants bone, muscle, and a certain amount of fat, and these are procured from the grasses, roots, and grain; from the former when fed alone, and from the two latter when mixed with hay or grass.

Horses, cattle, and sheep need hay to qualify the too watery nature of the roots, and the too condensed nutritiveness of the grain. Animals that are preparing for the shambles, require vegetable oils or fat, starch, sugar, or gum. The first is contained in great abundance in flax and cotton-seed, the sun-flower, and many other of the mucilaginous seeds. Indian corn is the most fattening grain. The potato contains the greatest proportion of starch, and the sugar-beet has large quantities of sugar, and both consequently are good for stall-feeding. The ripe sugar-cane is perhaps the most fattening of vegetables, if we except the oily seeds and grain. The Swedes turnip is a good food to commence with fattening cattle and sheep; but where great ripeness in animals is desired, they should be followed with beets, carrots or potatoes, and grain.

The table of the average composition of the different crops, which we subjoin from Johnston, shows the comparative qualities of various kinds of food, and it will be found a valuable reference for their nutritive and fattening qualities. He says, "In drawing up this table, I have adopted the proportions of gluten, for the most part, from Boussingault. Some of them, however, appear to be very doubtful. The proportions of fatty matter are also very uncertain. With a few exceptions, those above given have been taken from Sprengel, and they are, in general, stated considerably too low. It is an interesting fact, that the proportion of fatty matter in and immediately under the husk of the grains of corn, is generally much greater than in the substance of the corn itself. Thus I have found the pollard of wheat to yield more than twice as much oil as the fine flour obtained from the same sample of grain. The four portions separated by the miller from a superior sample of wheat grown in the neighborhood of Durham, gave of oil respectively: fine flour, 1·5 per cent.; pollard, 2·4; boxings, 3·6; and bran, 3·3 per cent. Dumas states that the husk of oats sometimes yields as much as five or six per cent of oil." The columns under starch, &c., and fatty matter, denote the value for respiration or sustaining life, and the fattening qualities; that under gluten, the capacity for yielding muscle and supporting labor; and saline matter indicates something of the proportions which are capable of being converted into bones.

	Water.	**Husk or woody fibre.**	**Starch, gum, and sugar.**	**Gluten, albumen, legumen, &c.**	**Fatty matter.**	**Saline matter**
Wheat,	16	15	55	10 to 15	2 to 4 J.	2·0
Barley,	15	15	60	12?	2·5 J.	2·0
Oats,	16	20	50	14·5?	5·6 J.	3·5
Rye,	12	10	60	14·5	3·0	1·0
Indian corn,	14	15?	50	12·0	5 to 9 D.	1·5
Buckwheat,	16?	25?	50	14·5	0·4?	1·5
Beans,	16	10	40	28·0	2 +	3·0
Peas,	13	8	50	24·0	2·8?	2·8
Potatoes,	75?	5?	12?	2·25	0·3	0·8 to 1
Turnips,	85	3	10	1·2	?	0·8 to 1
Carrots,	85	3	10	2·0	0·4	1·0
Meadow hay,	14	30	40	7·1	2 to 5 D.	5 to 10
Clover hay,	14	25	40	9·3	3·0	9
Pea straw,	10 to 15	25	45	12·3	1·5	5
Oat do.	12	45	35	1·3	0·8	6
Wheat do.	12 to 15	50	30	1·3	0·5	5
Barley do.	do.	50	30	1·3	0·8	5
Rye do.	do.	45	38	1·3	0·5	3
Indian corn do.	12	25	52	3·0	1·7	4

This table, it will be perceived, is far from settling the *precise* relative value of the different enumerated articles. An absolute, unchanging value can never be assumed of any one substance, as the quality of each must differ with the particular variety, the soil upon which it is grown, the character of the season, the manner of curing, and other circumstances. An approximate relative value is all that can be expected, and this we may hope ere long to obtain, from the spirit of analytical research, which is now developed and in successful progress. More especially do we need these investigations with American products, some of which are but partially cultivated in Europe, whence we derive most of our analyses. And many which are there reared, differ widely from those produced here, as these also differ from each other. What, for instance, is the character of meadow hay? We know that this varies as four to one, according to the particular kinds grown; and our Indian corn has certainly a less range than from five to nine.

The Changes in the Food of Animals

Potatoes, when first ripe, are estimated to be worth, for feeding purposes, nearly twice as much as when old; and the relative value of the different kinds, varies greatly at the same age and under similar conditions of growth. Perrault ascertained by careful experiment, that hay, clover, and lucerne lost much of their nutritive qualities by drying, and in lucerne this loss amounted to about thirty-five per cent. This is an important consideration in the feeding of green and dry forage. Oats are among the best feed, both for young and working animals; but it has been found that they are greatly improved for the latter, and perhaps for both, by allowing the new crop to remain till the latter part of winter, before feeding.

The improvement by steaming and cooking food, is seldom sufficiently appreciated. Food properly managed, can never be made worse by cooking for any stock; although it has not been considered so essential for working, and generally, for ruminating animals, as for swine, and such as were stall-feeding. But the alteration produced in cooking, by fitting it for a more ready assimilation, must, as a general rule, add much to the value of the food, and the rapid improvement of the animal.

The effect of slight fermentation, or souring the food, produces the same result. Animals accustomed to this acid food, will reject what is unprepared when they can get at the former; and we have no doubt, from our own experience, that there is a saving in thus preparing it, from 20 to 40 per cent. A mixture of food should be supplied to all animals. Like man, they tire of any constant aliment. For such, especially, as are fattening, and which it is desirable to mature with the greatest rapidity, a careful indulgence of their appetite should be studied. They should be provided with whatever they most crave, if it be adapted to the secretion of fat. Cutting, crushing, and grinding the food; cooking, souring, and mixing it, are each by themselves an improvement for feeding; and frequently two or more of these preparations combined, are of great utility in effecting the object proposed.

The Profit of Feeding

It is evident, that this consists in a valuable return from the animal of the food consumed. In the horse, this can only be received in labor or breeding; in the ox, from labor and flesh; in the cow, from the milk, the flesh, and her young. In the sheep, it may be returned in its fleece, its carcass, or its progeny; and in the swine only by its progeny and flesh. The manure we expect from all; and if this be not secured and judiciously used, few animals about the farm will be found to yield a satisfactory profit for their food and attention; though it is evident, it should form but a small part of the return looked for.

Animals are only profitable to the farmer, when they yield a daily income, as in their milk or labor; or annually, by their young or fleece; unless it be in a course of regular improvement, either in their ordinary growth or preparation for the butcher. The animal must consume a certain amount of food merely to keep up its stationary condition, and to supply the materials for waste, respiration, perspiration, and the evacuations. These must first be provided for in all cases, before the farmer can expect anything for the food.

Frequent observation has shown, that an ox will consume about two per cent, of his weight of hay per day, to maintain his condition. If put to moderate labor, an increase of this quantity to three per cent., will enable him to perform his work, and still maintain his flesh. If to be fattened, he requires about 4½ per cent of his weight daily, in nutritious food. A cow to remain stationary and give no milk, eats two per cent of her weight daily;

and if in milk, she will consume three per cent. If these statements are correct, which it is certain they are in principle, though they may not be entirely in degree, it will require the same food to keep three yoke of cattle in idleness, as two at work; and the food of every two that are idle, will nearly support one under the most rapid condition of fatting. Two cows may be kept in milk, with the same feed that will keep three without.

No practice is more impolitic, than barely to sustain the stock through the winter, or a part of the year, as is the case in too many instances, and allow them to improve only when turned on grass in summer. Besides subjecting them to the risk of disease, consequent upon their privation of food, nearly half the year is lost in their use, or in maturing them for profitable disposal; when if one-third of the stock had been sold, the remainder would have been kept in a rapidly improving condition, and at three years of age, they would probably be of equal value, as otherwise at five or six. It is true that breed has much to do with this rapid advancement, but breed is useless without food to develop and mature it.

Chapter 2

NEAT OR HORNED CATTLE

The value of our neat cattle exceeds that of any other of the domestic animals in the United States. They are as widely disseminated, and more generally useful. Like sheep and all our domestic brutes, they have been so long and so entirely subject to the control of man, that their original type is unknown. They have been allowed entire freedom from all human direction or restraint for hundreds of years, on the boundless pampas of South America, California, and elsewhere; but when permitted to resume that natural condition, by which both plants and animals approximate to the character of their original head, they have scarcely deviated in any respect, from the domestic herds from which they are descended. From this it may be inferred, that our present races do not differ, in any of their essential features and characteristics, from the original stock.

VARIOUS DOMESTIC BREEDS

Cultivation, feed, and climate, have much to do in determining the form, size, and character of cattle. In Lithuania, cattle attain an immense size, with but moderate pretensions to general excellence, while the Irish Kerry and Scotch Grampian cows but little exceed the largest sheep; yet the last are compact and well-made, and yield a good return for the food consumed. Every country, and almost every district, has its peculiar breeds, which by long association have become adapted to the food and circumstances of its position, and when found profitable, they should be exchanged for others, only after the most thorough trial of superior fitness for the particular location, in those proposed to be introduced.

More attention has been paid to the improvement of the various breeds of cattle in England, than in any other country; and it is there they have attained the greatest perfection in form and character for the various purposes to which they are devoted. We have derived, directly from Great Britain, not only the parent stock from which nearly all our cattle are descended, but also most of those fresh importations, to which we have looked for improvement on the present race of animals.

A few choice Dutch cattle, generally black and white, and of large size, good forms, and good milkers, with a decided tendency to fatten, have been occasionally introduced among us, but not in numbers sufficient to keep up a distinct breed; and in the hands of their importers, or immediate successors, their peculiar characteristics have soon become merged in those herds by which they were surrounded. Some few French and Spanish cattle, the descendants of those remote importations, made when the colonies of those kingdoms held possession of our northern, western, and southern frontiers, still exist in those sections; and although possessing no claims to particular superiority, at least in any that have come within our notice, yet they are so well acclimated, and adapted to their various localities, as to render it inexpedient to attempt supplanting them, except with such as are particularly meritorious.

Native Cattle

This is a favorite term with Americans, and comprehends everything in the country, excepting such as are of a pure and distinct breed. It embraces some of the best, some of the worst, and some of almost every variety, shape, color, and character of the bovine race. The designation has no farther meaning, than that they are indigenous to the soil, and do not belong to any well-defined or distinct variety.

The best native cattle of the Union are undoubtedly to be found in the Northeastern states. Most of the early emigrant cattle in that section were from the southern part of England, where the Devon cattle abound; and though not at the present time bearing a close resemblance to that breed, unless it has been impressed upon them by more recent importations, yet a large number have that general approximation in character, features, and color, which entitles them to claim a near kindred with one of the choicest cultivated breeds. They have the same symmetry, but not in general the excessive delicacy of form, which characterizes the Devons; the same intelligence, activity, and vigor in the working cattle, and the same tendency to fattening; but they are usually better for the dairy than their imported ancestors. Some valuable intermixtures have occasionally been made among them. Among these, there have been many brindled cattle widely disseminated, of great merit as workers, and not often surpassed for the dairy and shambles.

The Herefords have in a few instances been introduced among the eastern cattle, and apparently with great improvement. The importation made by Admiral Coffin, of four choice Hereford bulls and cows, which were presented to the State Agricultural Society of Massachusetts, nearly thirty years since, is especially to be mentioned, as resulting in decided benefit wherever they were disseminated. Some of the old Yorkshire, or as they are sometimes styled, the long-horned Durhams, have been introduced, though these have been isolated individuals and never perpetuated as a separate breed. A few small importations have been made of the Short Horns and Ayrshires, but neither of these have been bred in the New England states in distinct herds, to any extent.

Their *native breed* has hitherto, and generally with good reason, possessed claims on the attention of their owners, which, with some slight exceptions, it has not been in the power of any rivals to supplant. With entire adaptedness to the soil, climate, and wants of the farmer, an originally good stock has, in frequent instances, been carefully fostered, and the breeding animals selected with a strict reference to their fitness for perpetuating the most desirable qualities. As a consequence of this intelligent and persevering policy, widely, but not universally pursued, they have a race of cattle, though possessing considerable diversity of size and color, yet coinciding in a remarkable degree in the possession of those utilitarian features, which so justly commend them to our admiration.

In proceeding southwestwardly through New York, New Jersey, and elsewhere, we shall find in this branch of stock, a greater diversity and less uniform excellence; though they have extensive numbers of valuable animals. Here and there will be found a choice collection of some favorite foreign breed, which emigrants have brought from their native home, as did the Pagan colonists their penates or household gods; the cherished associates of early days, and the only relics of their father-land. Such are an occasional small herd of polled or hornless cattle, originally derived from Suffolk or Galloway, excellent both for the dairy and shambles; the Kyloe, or West Highland, (Scottish,) a hardy animal, unrivalled for beef; the Welsh runt; the Irish cattle; the crumpled-horn Alderney, and some others.

The Devon

Is among the oldest distinctly cultivated breeds in this country, as it undoubtedly is of England, and probably it is the most universal favorite. This popularity is well deserved, and it is based upon several substantial considerations. They are beautifully formed, possessing excessive fineness and symmetry of frame, yet with sufficient bone and muscle to render them perfectly hardy; and they are among the most vigorous and active of working cattle. They have great uniformity of appearance in every feature, size, shape, horns, and color. The cows and bulls appear small, but the ox is much larger; and both he and the dam, on cutting up, are found to weigh much beyond the estimates which an eye accustomed only to ordinary breeds, would have assigned to them. The flesh is finely marbled or interspersed with alternate fat and lean, and is of superior quality and flavor.

The cows invariably yield milk of great richness, and when appropriately bred, none surpass them for the quantity of butter and cheese it yields. Mr. Bloomfield, the manager of the late Lord Leicester's estate at Holkham, has, by careful attention, somewhat increased the size, without impairing the beauty of their form, and so successful has he been in developing their milking properties, that his average product of butter from each cow, is 4 lbs. per week for the whole year. He has challenged England to milk an equal number of cows of any breed, against 40 pure Devons, to be selected out of his own herd, without as yet having found a competitor. Although this is not a test of their merits, and by no means decides their superiority, yet it shows the great confidence reposed in them by their owner. The Devon ox, under six years old, has come up to a nett dead weight of 1,593 lbs.; and at three years and seven months, to 1,316 lbs., with 160 lbs. of rough tallow.

Description. The Devon is of medium size, and so symmetrical, as to appear small. The color is invariably a deep mahogany red, with usually a white udder and strip under the belly; and the tuft at the end of the tail is red while they are calves, but white in the older animal. The head is small, broad in the forehead, and somewhat indented. The muzzle is delicate, and both the nose and the rings around the eye, in the pure breed, are invariably of a bright, clear orange. The cheeks and face are thin and fleshless; the horns clear, smooth, and of a yellowish white, handsomely curved upward. The neck is small and delicate at its junction with the head, but is well expanded in its attachment to the breast and shoulders. The last has the true slant for activity and strength, in which it excels all other breeds of equal weight. The barrel is round and deep, with a projecting brisket. The back is broad and level; the flank full; hips wide; the rumps long; the quarters well developed, and capable of holding a great quantity of the most valuable meat. The tail is on a level with the back, and gracefully tapers like a drum-stick, to the tuft on the end. The legs are of peculiar delicacy and fineness, yet possess great strength. The skin is of medium thickness, of a rich orange hue, pliable to the touch, and covered with a thick coating of fine, soft, curly hair. The Devon is intelligent, gentle, and tractable; is good for milk, and unsurpassed for the yoke and for fattening. No animal is better suited to our scanty or luxuriant hill pastures than the Devon, and none make a better return for the attention and food received. They ensure a rapid improvement when mixed with other cattle, imparting their color and characteristics in an eminent degree. Several importations have been made into this country within the last 30 years, of the choicest animals, and though not yet numerous in the United States, we possess some of the best specimens that exist.

The Short Horns, or Durhams

Are decidedly the most showy among the cattle species. They are of all colors between a full, deep red, and a pure creamy white; but generally have both intermixed in larger or smaller patches, or intimately blended in a beautiful roan. Black, brown, or brindled, are colors not recognised among pure-bred Short Horns. Their form is well-spread, symmetrical, and imposing, and capable of sustaining a large weight of valuable carcass. The horn was originally branching and turned upward, but now frequently has a downward tendency, with the tips pointing towards each other. They are light, and comparatively short; clear, highly polished, and waxy. The head is finely formed, with a longer face but not so fine a muzzle as the Devon. The neck is delicately formed without dewlap, the brisket projecting; and the great depth and width of the chest giving short, well-spread fore-legs. The crops are good; back and loin broad and flat; ribs projecting; deep flank and twist; tail well set up, strong at the roots and tapering. They have a thick covering of soft hair, and are mellow to the touch, technically termed, *handling well*. They mature early and rapidly for the quantity of food consumed, yielding largely of good beef with little offal. As a breed, they are excellent milkers; though some families of the Short Horns surpass others in this quality. They are inferior to the Devons, in their value as working oxen, and in the richness of their milk.

Figure 1. A Short-Horn Bull.

The Short Horns are assigned a high antiquity, by the oldest breeders in the counties of Durham and Yorkshire, England, the place of their origin, and for a long time, of their almost exclusive breeding. From the marked and decided improvement which they stamp upon other animals, they are evidently an ancient breed, though much the juniors of the Devon and Hereford. Their highly artificial style, form, and character, are unquestionably the work of deeply studied and long-continued art; and to the same degree that they have been moulded in unresisting compliance with the dictation of their intelligent breeders, have they departed from that light and more agile form of the Devon, which conclusively and beyond the possibility of contradiction, marks the more primitive race.

The Importation of Short Horns into this Country

This is claimed to have been previous to 1783. They are the reputed ancestors of many choice animals existing in Virginia, in the latter part of the last century, and which were known as the *milk breed*; and some of these, with others termed the *beef breed*, were taken into Kentucky by Mr. Patton, as early as 1797, and their descendants, a valuable race of animals, were much disseminated in the West, and known as the *Patton stock*.

The first authentic importations we have recorded, are those of Mr. Heaton, into Westchester, N. Y., in 1791 and '96, from the valuable herds of Messrs. Culley and Colling, which consisted of several choice bulls and cows. These were for many years bred pure, and their progeny was widely scattered. (*American Herd Book.*) They were also imparted into New York, by Mr. Cox, in 1816; by Mr. Bullock, in 1822; by the late Hon. S. Van Rensselaer in 1823; and immediately after, by Mr. Charles Henry Hall, of Harlem. Some small importations were made into Massachusetts between 1817 and '25, by several enterprising agriculturists, Messrs. Coolidge, Williams, and others; into Connecticut by Mr. Hall and others; into Pennsylvania by Mr. Powell; and into Ohio and some other states, by various individuals early in the present century.

Figure 2. Short-Horned Cow.

Since the first importations, larger accessions from the best English herds have been frequently made; and with the nice regard for pedigrees which the introduction of the herd book, and careful purity in breeding has produced, the Short Horns have become the most extensive pure-bred family of cattle in the United States.

During the speculative times of 1835 to 1840, they brought high prices, frequently from $500 to $1000, and sometimes more. The following years of financial embarrassment, reduced their market price below their intrinsic value; but the tide is again turning, and they are now in demand, but still at prices far below their utility and merits. They have from the first, been favorites in the rich, corn valleys of the West, their early maturity and great weight giving them a preference over any other breed. The only drawback to this partiality, is their inability, from their form and weight, to reach remote eastern markets in good condition; an objection now in a great measure remedied, by the recent remission of duties on foreign beef in the English market, which makes them of nearly equal value where fed, to pack for exportation. On light lands and scanty pastures, they will probably never be largely introduced. All heavy animals require full forage within a limited compass, so as to fill their stomachs at once, and quietly compose themselves to their digestion.

The weights reached by the Short Horns in England, as given by Mr. Berry, have been enormous. Two oxen, six years old, weighed nett, 1820 lbs. each. A heifer of three years, and fed on grass and hay alone, weighed 1260 lbs. A four-year-old steer, fed on hay and turnips only, dressed 1890 lbs. A cow reached the prodigious weight of 1778 lbs. A heifer, running with her dam, and on pasture alone, weighed at seven months, 476 lbs. An ox, seven years old, weighed 2362 lbs. From their comparatively small numbers in this country, most of them have been retained for breeders; few, as yet, have been fattened, and such only as were decidedly inferior. Such animals as have been extensively produced by crossing this breed upon our former stocks, have given evidence of great and decided improvement; and the Short Horns, and their grade descendants are destined, at no distant day, to occupy a large portion of the richest feeding grounds in the United States.

Figure 3. Hereford Cow.

Herefords

This is the only remaining pure breed, which has hitherto occupied the attention of graziers in this country. Like the Devons, they are supposed to be one of the most ancient races of British cattle. Marshall gives the following description. "The countenance pleasant, cheerful, open; the forehead broad; eye full and lively; horns bright, taper, and spreading; head small; chap lean; neck long and tapering; chest deep; bosom broad, and projecting forward; shoulder-bone thin, flat, no way protuberant in bone (?) but full and mellow in flesh; chest full; loin broad; hips standing wide, and level with the chine; quarters long, and wide at the neck; rump even with the level of the back, and not drooping, nor standing high and sharp above the quarters; tail slender and neatly haired; barrel round and roomy; the carcass throughout deep and well spread; ribs broad, standing flat and close on the outer surface, forming a smooth, even barrel, the hindmost large and full of length; round-bone small, snug, and not prominent; thigh clean, and regularly tapering; legs upright and short; bone below the knee and hock small; feet of middle size; flank large: flesh everywhere mellow, soft, and yielding pleasantly to the touch, especially on the chine, the shoulder, and the ribs; hide mellow, supple, of a middle thickness, and loose on the neck and huckle; coat neatly haired, bright and silky; color, a middle red, with a bald face characteristic of the true Herefordshire breed."

Figure 4. Hereford Bull.

Youatt further describes them as follows: "They are usually of a darker red; some of them are brown, and even yellow, and a few are brindled; but they are principally distinguished by their white faces, throats, and bellies. In a few the white extends to the shoulders. The old Herefords were brown or red-brown, with not a spot of white about them. It is only within the last fifty or sixty years that it has been the fashion to breed for white faces. Whatever may be thought of the change of color, the present breed is certainly far superior to the old one. The hide is considerably thicker than that of the Devon, and the beasts are more hardy. Compared with the Devons, they are shorter in the leg, and also in the carcass; higher, and broader, and heavier in the chine; rounder and wider across the hips, and better covered with fat; the thigh fuller and more muscular, and the shoulders larger and coarser.

They are not now much used for husbandry, although their form adapts them for the heavier work; and they have all the honesty and docility of the Devon ox, and greater strength, if not his activity. The Herefordshire ox fattens speedily at a very early age, and it is therefore more advantageous to the farmer, and perhaps to the country, that he should go to market at three years old, than be kept longer as a beast of draught.

They are not as good milkers as the Devons. This is so generally acknowledged, that while there are many dairies of Devon cows in various parts of the country, a dairy of Herefords is rarely to be found. To compensate for this, they are even more kindly feeders than the Devons. Their beef may be objected to by some as being occasionally a little too large in the bone, and the fore-quarters being coarse and heavy; but the meat of the best pieces is often very fine-grained and beautifully marbled. There are few cattle more prized in the market than the genuine Herefords."

There have been several importations of the Herefords into the United States, which by crossing with our native cattle, have done great good; but with the exception of a few fine animals at the South, we are not aware of their being kept in a state of purity, till the importation of the splendid herd. within the last six years, by Messrs. Corning and Sotham of Albany, N. Y. These Herefords are among the very best which England can produce, and come up fully to the description of the choicest of the breed. Mr. Sotham, after an experience of several years, is satisfied with the cows for the dairy; and he has given very favorable published statements of the results of their milking qualities, from which it may be properly inferred, that Youatt drew his estimates from some herds which were quite indifferent in this property. They are peculiarly the grazier's animal, as they improve rapidly and mature early on medium feed. They are excelled for the yoke, if at all, only by the Devons, which, in some features, they strongly resemble. Both are probably divergent branches of the same original stock.

The Ayrshire

Is a breed that has been much sought after of late years, from their reputation for fine dairy qualities. The milk is good both in quantity and quality, yielding, according to a recent statement of Mr. Tennant, of Scotland, who owns a large herd, fifteen quarts per day during the best of the season, twelve of which made a pound of butter. The product of the latter averages about 170 pounds per annum to each cow. Another authority says, on the best low-land pasture, a good cow yields nearly 4000 quarts per year. This is a large quantity, and implies good cows and extra feed.

Mr. Cushing, of Massachusetts, who imported several select animals, without regard to their cost, informed us, after three or four years' trial, that he did not perceive any superiority in them, over the good native cows of that state, for dairy purposes. A large number have been imported in detached parcels, and scattered through the country. They are good animals, but seem to combine no valuable properties in a higher degree than are to be found in our own good cattle, and especially such as are produced from a cross of the Short Horn bull of a good milking family, on our native cows. They are evidently a recent breed, and do not therefore possess that uniformity of appearance and quality, which attaches to one of long cultivation.

Mr. Aiton, of Scotland, gives the following account of them. "The dairy breed of Scotland have been formed chiefly by skilful management, within the last 50 years; and they are still improving and extending to other countries. Till after 1770, the cows in Cunningham were small, ill-fed, ill-shaped, and gave but little milk. Some cows of a larger breed and of a brown and white color, were about that time brought to Ayrshire from Teeswater, and from Holland, by some of the patriotic noblemen of Ayrshire; and these being put on good pasture, yielded more milk than the native breed, and their calves were much sought after by the farmers."

We may fairly infer from the foregoing, which is deemed indisputable authority; from the locality of their origin, in the neighborhood of the Short Horns; and from their general resemblance, both externally and in their general characteristics to the grade animals, that they owe their principal excellence to this long-established breed.

Management of Calves

The safest and least troublesome manner of raising calves, is at the udder of the dam; and whenever the milk is converted into butter and cheese, we believe this to be the most economical. The milk of one good cow is sufficient, with a run of fresh, sweet pasture, to the feeding of two calves at the same time; and if we allow the calves to arrive at three or four months of age before weaning, we may safely estimate, that one good cow will yield a quantity of milk in one season, fully equivalent to bringing up four calves to a weaning age.

By keeping the calf on the fresh milk, whether he take it directly from the udder, or warm from the pail, all risk of disordered bowels is avoided. The milk is precisely adapted to the perfect health and thrift of the young, and whenever we substitute for it any other food, we must watch carefully that not the slightest mismanagement produces disorder, lest more is lost by disease or want of improvement, than is gained by the milk of which they are robbed.

The first milk of the cow after calving, is slightly purgative, which is essential to cleanse the stomach of the calf. It is, moreover, perfectly worthless for two or three days, for any other purpose except for swine. The calf will seldom take all the milk at first, and whatever is left in the bag should be thoroughly removed by the hand. If the calf is destined for the butcher, he must have all the milk he wants for at least six weeks, and eight or ten is better; and if the cow does not furnish enough, he ought to be fed gruel or linseed tea. He must be closely confined in a snug, but clean and airy stable, and the darker this is, and the more quiet he is kept, the more readily he will fatten.

If designed to be reared, the safest and least troublesome method, is to keep the calf on new milk. If saving the milk be an object, it is still doubtful whether it is not better that he should have a part of it fresh from the cow, and depend for his remaining food on a good grass or clover pasture, meal, or roots.

Some farmers never allow the calf to approach the dam, but take it when first dropped, and put a handful of salt in its mouth, which is daily repeated till he is put to grass. This has a purgative effect, similar to the first milk. Flaxseed is then prepared, by boiling a pint in four to six quarts of water, and diluted with hay tea till it is rather thicker than milk, and fed at blood heat.

Hay tea is made, by boiling a pound of sweet, well-cured clover, in one and a half gallons of clean water.

As the calf becomes older, oat, barley, rye, or Indian meal may be scalded and added to the flaxseed.

When the skim-milk is of little consequence, a better way is to withdraw him from the cow after three or four days, then scald the milk, adding a little oat meal, and cool to the natural temperature of the milk, and feed it. Oats, either crushed or ground, is the best and safest grain for all young stock. The milk should not stand more than half a day before feeding to young calves. As they advance in age, it may be fed rather older, but should never be allowed to become sour; nor should it ever be fed cold. Connected with this feed, should be a good range of short, sweet pasture, and shelter against both sun and storms. If expedient, at about 10 weeks old, he may be safely weaned, but four months' nursing is better for the calf.

If allowed too much milk for several months, it is injurious to the future development of the young. It does not distend the stomach properly, nor call into use its ruminating habits. Calves thus brought up, have often proved light-bellied, indifferent feeders, and decidedly inferior animals. When the calf is removed from the cow, they should be effectually separated from sight and hearing, as recognition creates uneasiness, and is an impediment to thrift in both.

If there be any deficiency of suitable pasture for the calf, a small rack and trough should be placed under the shed in his range, and fine hay put in the former, and wheat bran or oat meal with a little salt in the latter.

Diseases and Remedies

For *disordered bowels*, mix 2 dr. rhubarb, 2 oz. castor oil, and ½ dr. ginger, with a little warm milk or gruel; or give 2 oz. castor oil alone; or 3 oz. of Epsom salts.

For *scours* and *diarrhœa*, a homely remedy is, to administer half a pint of cider, with an equal quantity of blood drawn from the calf's neck.

Or, add a little rennet to its food.

A good remedy is, 1 oz. powdered canella bark; 1 oz. laudanum; 4 oz. prepared chalk; and one pint water. Mix together, and give a wine-glass full or more, according to the size of the calf, three times a day.

Costiveness is removed by giving pork broth.

Or, give 3 to 4 oz. Epsom salts, dissolved in 3 pints of water, injected into the stomach; and repeat part of this dose every 3 or 4 hours, till the desired effect is produced.

Calves, like all young stock, should be allowed to change their feed gradually, from new milk to skimmed, or from the latter to other food. Their stomachs are delicate, and need gentle, moderate changes, when necessary to make them at all. Much depends on the care and attention they receive. It is well to have a little resin within its reach.

A comfortable shelter, with a dry, warm bed, suitable food, regularly given three times a day, at blood heat, and keeping the stomach in proper order, will do much to bring them forward rapidly, and with a small expenditure of food.

The calf requires to be supplied through the winter with an abundance of fine, sweet hay and roots, the latter either chopped or mashed by a roller, with the addition of a trifle of meal or oats, and a full supply of salt and pure water.

When there are larger animals on the premises, the calves ought to be kept by themselves. They should be sustained on their winter feed through the following spring, until the grass furnishes a good bite on a well-compacted sod. The change from hay to grass must be gradual, unless the latter is considerably matured. The extreme relaxation of the bowels from the sudden change, frequently produces excessive purging. A slight and temporary relax from the early spring grass, is not objectionable.

Breeding

The young animals should never be put to breeding under 15 months old, so as to bring their first calf at two years old nor then, unless they have large size and good feed. Much depends on the progress towards maturity, and the supply of food in selecting the proper time for breeding. Some are as ready for this at a year and a half as others are at three. Early breeding gives delicacy and symmetry to the form of the heifer, but it checks its growth; and when it is found to put her back too much, she may be allowed to rest for a few months, or even a year, to brings her up to the desired standard. These remarks apply principally to choice breeders, or as they are sometimes termed, *fancy stock*. For ordinary milch cows which have been moderately fed, three years is a proper age to come in, after which they must be milked as regularly, and as late before drying as possible.

Breaking Steers

Should be commenced when two or three years old. Some begin with the calf, accustoming him to a light yoke and occasional training. This practice will do as a pastime for trustworthy boys, as it makes them gentle and manageable afterwards, but is hardly worth a man's time. If always carefully handled when young, they will be found tractable.

They should at first be placed behind a pair of well-broke cattle, nor should they be put to hard labor until quite grown, strong, and perfectly accustomed to the yoke. If properly managed, cattle may be trained with all the docility, intelligence, and much of the activity of the horse. That they are not, is more frequently the fault of their masters.

Management of Oxen

To procure perfect working cattle, it is necessary to begin with the proper breed. Many parts of the country furnish such as are well suited to this purpose. A strong dash of Devon or Hereford blood is desirable, when it needs to be improved. A well-formed, compact, muscular body; clean, sinewy limbs; strong, dense bones; large, well-formed joints, with a mild expressive eye, are essential for good working oxen.

After breaking, they must be led along gently, and taught before they are required to perform their task; and never put to a load which they cannot readily move, nor dulled by prolonging exertion beyond that period when it becomes irksome. A generous diet is necessary, to keep up the spirit and ability of cattle, when there is hard work to be done. The horse and mule are fed with their daily rations of grain when at hard service, and if the spirit of the ox is to be maintained, he should be equally well fed, when as fully employed. Great and permanent injury is the result of niggardly feeding and severe toil, exacted from the uncomplaining animal. His strength declines, his spirit flags, and if this treatment be continued, he rapidly becomes the stupid, moping brute, which is shown off in degrading contrast with the more spirited horse, that performs, it may be, one half the labor, on twice his rations.

The ox should be as little abused by threats and whipping, as by stinted feed and overtasked labor. Loud and repeated hallooing, or the severe use of the lash, is as impolitic as it is cruel and disgraceful. We never witness this barbarity without wishing the brutes could change places, long enough at least to teach the biped that humanity by his own sufferings which his reason and sensibility have failed to inspire. Clear and intelligible, yet low and gentle words are all that are necessary to guide the well-trained, spirited ox. The stick, or whip, is needed rather to indicate the precise movement desired, than as a stimulant or means of punishment. The ox understands a moderate tone more perfectly than a boisterous one, for all sounds become indistinct as they increase.

It is of great advantage to have oxen well trained to *backing*. They may soon be taught, by beginning with an empty cart on a descent; then on a level; then with an increasing load, or uphill, till the cattle will back nearly the same load they will draw.

Some oxen have a bad trick of *hauling* or *crowding*. Changing to opposite sides, longer or shorter yokes, and more than all, gentle treatment, are the only remedies, and those not unfrequently fail. Cattle will seldom contract this habit, in the hands of a judicious, careful driver. *The yokes* should be carefully made, and set easy, and the bows fitted to the necks and properly attached to the yoke. Cattle are liable to sore necks if used in a storm; and when subject to this exposure, they must be well rubbed with grease, where the yoke chafes them, and respite from work should be allowed till the necks heal.

Management of Bulls or Vicious Animals

If inclined to be vicious, the bulls should have rings thrust through the cartilage of their nose when young. They are to be found at the agricultural warehouses; and are made of round iron, three-eighths of an inch diameter, with a joint in one side to open, and when thrust through the nose, are fastened in a moment, by a rivet previously prepared.

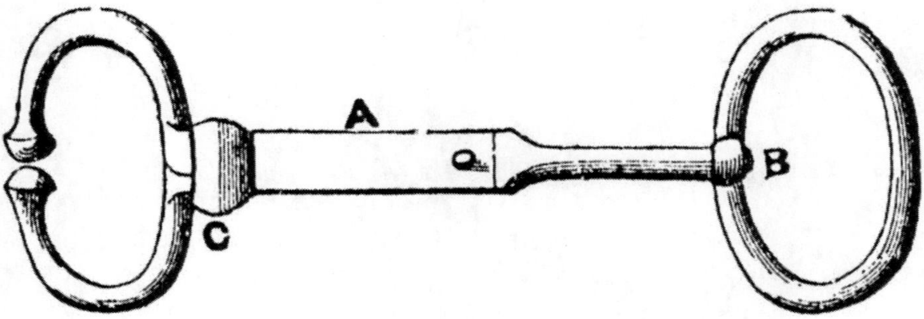

Figure 5. Cattle-Handler.

Figure 5 is a *cattle-handler*, consisting of a bar of iron A, eight inches long, with a ring for a man's hand, which turns on a swivel at B, and at the other end is a pair of calliper-shaped legs, one stationary, and the other opening on a joint. The fixed leg is inserted against one side of the nostril, and the other is pressed upon the opposite side, and there fastened by a slide, C, when the animal is firmly held for administering medicine or performing any operation.

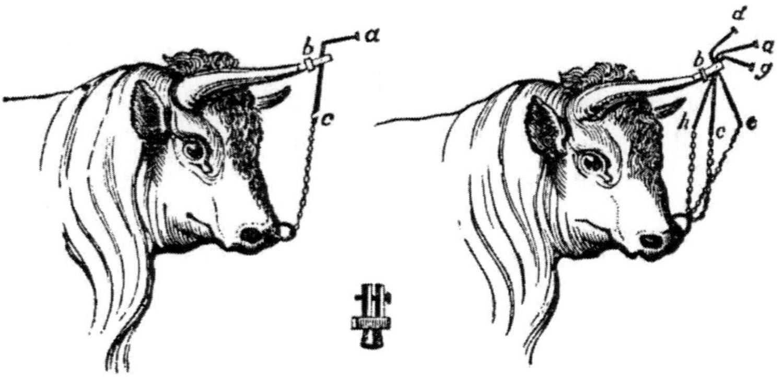

Figure 6 and Figure 7. For taming savage Animals.

Figures 6 and 7, for *taming a bull*; b, in Figure 6, is a cap screwed on to the tip of the horn; a c, an iron rod hanging on a pivot in the cap, with a chain reaching to the ring in the nose. The effect of his attempting to hook, is illustrated by the various positions of the chain in Figure 7. If the rod at a, is pushed in either direction, it jerks up the nose in a manner that cures him of his inclination.

Figure 8. Cattle-Tie.

Figure 8, shows a *cattle-tie.*—This is a much more convenient and comfortable mode of fastening cattle in the stable, than the common stanchions.

The Proper Time for Turning off Cattle

This must depend on their previous feeding and management, the breed, and the purposes required. The improved breeds and many of their crosses, will mature for the butcher as fully at three or four, as inferior cattle at five to seven years old. If pushed rapidly with proper food, they will of course be *ripe* much sooner than if stinted. When cattle have to be purchased for work, or cows for the dairy, it becomes an object to keep them as long as they can be made profitable, and yet be turned off for fattening at a fair price. We have seen active and spirited oxen in the yoke at 16 or 17; but they seldom do as well after 12 or even 10 years. Old cattle are liable to more diseases than young; are less hardy; and they recover more slowly when exposed to scanty feed or hard usage. They also fatten with more difficulty, and their meat is inferior. When they can be sold advantageously to the feeder, and replaced without inconvenience, it is found to be most profitable to turn them off at seven or eight years. They will by that time have attained full maturity; they will feed rapidly, and make the largest amount of good beef. If there are extraordinary milkers among the cows, or superior workers among the oxen, it is better to keep them as long as they maintain their full vigor.

Fattening Cattle

Such as are designed for the shambles the ensuing fall or winter, may be allowed to do their spring's labor; or if cows, they may be milked into summer after calving, or go farrow during the previous year. They should early be put on the best summer feed, and it is better to be occasionally changed, to give variety and freshness, and keep the animal in good appetite. Let the fattening animals have the best, and after they have cropped it a while, give them a fresh field; and the other animals or sheep can follow and clear off the remaining herbage, preparatory to shutting it up for a new growth. Some prefer an extensive range of rich feed, which is unchanged throughout the season; and when it is not necessary to divide the pasture with the other animals, this is a good practice.

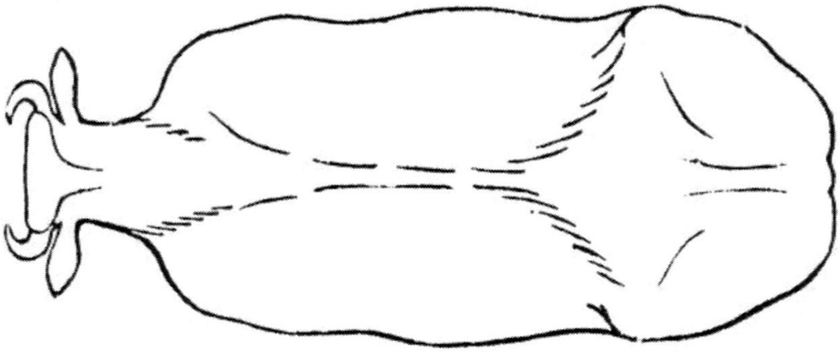

Figure 9.

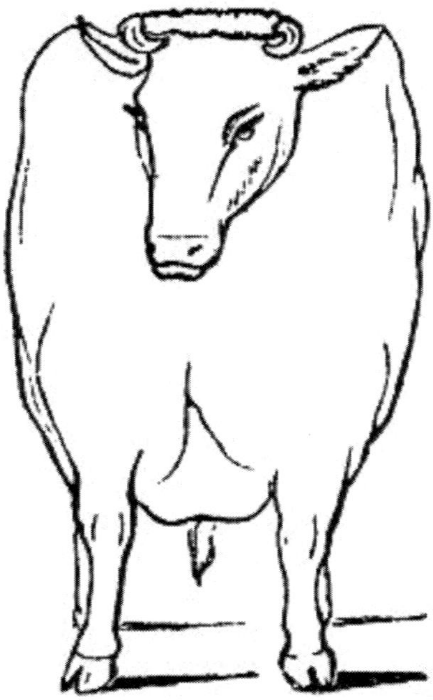

Figure 10.

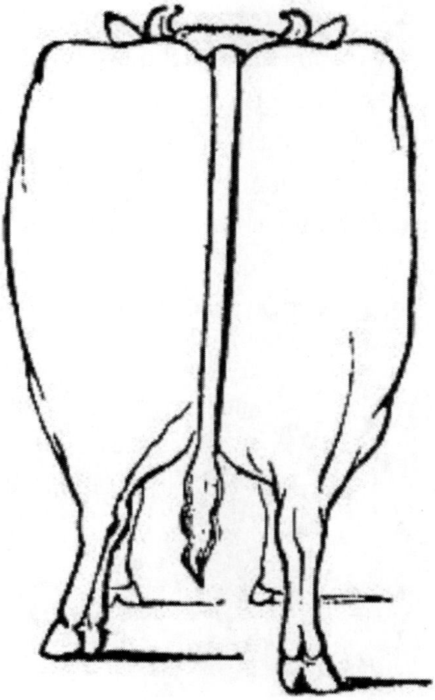

Figure 11.

Three cuts of improved forms, Nos. 9, 10, and 11. The above cuts illustrate the forms which the most improved beef-cattle should possess.

The Selection of Animals for Stall Fattening

This is a nice point, and none without a practised eye and touch, can choose such as will make the best return for the food consumed. The characteristics of choice animals, heretofore enumerated, are particularly essential in those intended for profitable fattening. But the most important of all, is that firm mellowness, and quick elasticity of touch, which unerringly mark the kindly feeder and profitable bullock. When other means for ascertaining fail, it is a safe rule to select the best-conditioned animals, out of a herd of grass-fed; for if all were of equal flesh and health, when turned out, those which have thriven most on their summer pasture, will generally fatten quickest on their fall and winter keep. Only the best should be selected. The remainder, after consuming the coarser forage, may be at once disposed of for early use. From repeated trials, it is found that the carcass of stall-fed animals will barely return the value of the materials consumed, and their manure is generally the only compensation for the time and attention bestowed. None but choice, thrifty beasts will pay for their food and attention, and all others will make their best returns, by an immediate disposal, after the surplus fodder is gone.

Figure 12. Points of Cattle Illustrated.

Explanation.—A, forehead; B, face; C, cheek; D, muzzle; E, neck; F, neck vein; G, shoulder point; H, arm; I, shank; J, gambril, or hock; K, elbow; L, brisket, bosom, or breast; M, shoulder; N, crops; O, loin; P, hip, hucks, hocks, or huckles; Q, crupper bone, or sacrum; R, rump, or pin-bone; S, round bone, thurl, or whirl; T, buttock; U, thigh, or gaskit; V, flank; W, plates; X, back, or chine; Y, throat; Z, chest.

Stall-Feeding

This ought to be commenced early in the season. An ox may be fed in a box-stall, or if accustomed to a mate, they do better by tying together with sufficient room, yet not so near as to allow of injuring each other. The building should be warm, but not hot; well ventilated, yet having no current of cold air passing through; and as dark as possible. The stall ought to be kept clean and dry, and a deep bed of clean straw is of decided advantage.

The ox should be first fed the inferior and most perishable roots with his grain and dry forage, and his food should be gradually increased in richness, as he advances towards maturity. The food and water should be given three times a day, from thoroughly cleaned mangers or troughs. The animal likes a change of food, in which he should be indulged as often as may be necessary. If he refuses his food, a temporary privation, or variety is essential. When the food is changed, he should be moderately fed at first, till he becomes accustomed to it, as there is otherwise danger of cloying, which is always injurious. The moment the animal has done feeding, the remainder of the food ought to be at once removed. He then lies down, and if undisturbed, rests quietly till the proper hour induces him again to look for his accustomed rations. Regularity in the time of feeding, is of the utmost consequence. An animal soon becomes habituated to a certain hour, and if it be delayed beyond this, he is restless and impatient, which are serious obstacles to speedy fattening.

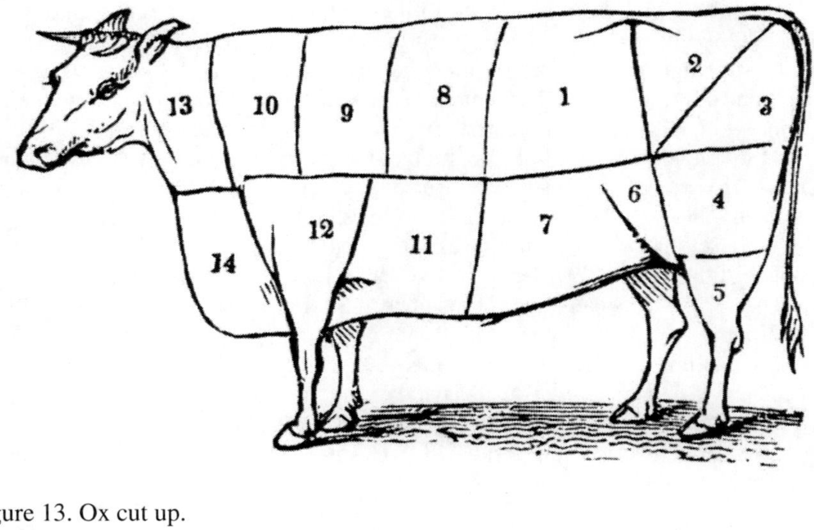

Figure 13. Ox cut up.

Figure 13—Shows the London method of cutting up the carcass—Figure 1, is the loin; 2, rump; 3, aitch or adz-bone; 4, buttock; 5, hock; 6, thick flank; 7, thin flank; 8, fore-rib; 9, middle rib; 10, cuck-rib; 11, brisket; 12, leg of mutton piece; 13, clod or neck; 14, brisket.

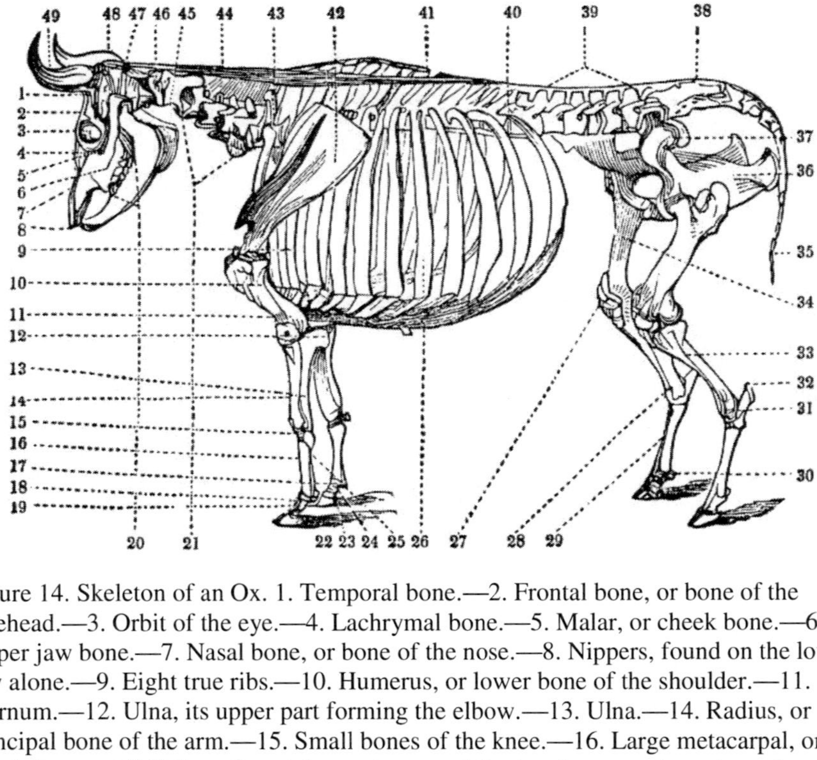

Figure 14. Skeleton of an Ox. 1. Temporal bone.—2. Frontal bone, or bone of the forehead.—3. Orbit of the eye.—4. Lachrymal bone.—5. Malar, or cheek bone.—6. Upper jaw bone.—7. Nasal bone, or bone of the nose.—8. Nippers, found on the lower jaw alone.—9. Eight true ribs.—10. Humerus, or lower bone of the shoulder.—11. Sternum.—12. Ulna, its upper part forming the elbow.—13. Ulna.—14. Radius, or principal bone of the arm.—15. Small bones of the knee.—16. Large metacarpal, or shank bone.—17. Bifurcation at the pasterns, and the two larger pasterns to each foot.—18. Sesamoid bones.—19. Bifurcation of the pasterns.—20. Lower jaw and the grinders.—21. Vertebræ, or bones of the neck.—22. Navicular bones.—23. Two coffin bones to each foot.—24. Two smaller pasterns to each foot.—25. Smaller or splint-bone.—26. False ribs, with their cartilages.—27. Patella, or bone of the knee.—28. Small bones of the hock.—29. Metatarsals, or larger bones of the hind leg.—30. Pasterns and feet.—31. Small bones of the hock.—32. Point of the hock.—33. Tibia, or proper leg-bone.—34. Thigh-bone.—35. Bones of the tail.—36, 37. Haunch and pelvis.—38. Sacrum.—39. Bones of the loins.—40. Bones of the back—41. Ligament of the neck and its attachments.—42. Scapula, or shoulder-blade.—43. Bones of the back.—44. Ligament of the neck.—45. Dentata.—46. Atlas.—47. Occipital bone, deeply depressed below the crest or ridge of the head—48. Parietal bone, low in the temporal fossa.—49. Horns, being processes or continuations of the frontal bone.

Diseases in Cattle

Hoven, or Swelling of the Paunch

Is a temporary ailment, caused by eating too freely of fresh and generally wet clover, or other succulent food. The animal gorges the first stomach with so much food, that its contents cannot be expelled. Inflammation of the membrane takes place, and decomposition of the food soon follows. This is known by the distension of the paunch, and difficulty of breathing, and unless speedily relieved, suffocation and death will ensue. Both sheep and cattle are subject to it.

Remedies.[1]—In its early stages, when not too severe, it has been removed by administering some one of the following remedies.

A pint of gin poured down the throat.

From one to two pints of lamp or other oil.

Strong brine.

New milk with one-fifth its bulk of tar mixed.

An egg-shell full of tar forced down the throat, followed by a second, if the first fails.

A tablespoonful of volatile spirit of ammonia, diluted with water.

A wine-glass full of powder, mixed with cold lard and forced in balls into the stomach.

A teaspoonful of unslaked lime dissolved in a pint of warm water, shaken and given immediately.

A pint of tolerably strong lye.

The Proper Mode of Giving the Above Remedies

Is for a person to hold the horn and cartilage of the nose, while another seizes and draws out the tongue as far as possible, when the medicine is

---

[1] Besides his own experience, the writer has drawn from the N. E. Farmer, the Albany Cultivator, the American Agriculturist, and other reliable American and English works, some of the remedies for diseases herein mentioned.

thrust below the root of the tongue. If liquid, it must be inserted by the use of a bottle.

The probang is used when the former remedies are ineffectual. This consists of a tarred rope, or a flexible whip-stalk, three-fourths of an inch in diameter, with a swab or bulbous end. Two persons hold the head of the animal, so as to keep the mouth in a line with the throat, while a third forces it into the stomach, when the gas finds a passage out. A stiff leather tube with a lead nozzle pierced with holes, is best for insertion, through which the gas will readily escape.

Some one of the above purgatives should be given after the bloat has subsided, and careful feeding for some days must be observed.

Light gruels are best for allaying inflammation, and restoring the tone of the stomach.

When no other means are available, the paunch may be tapped with a sharp penknife, plunging it 1½ inches forward of the hip bone, towards the last rib in the left side. If the hole fills up, put in a large goose-quill tube, which to prevent slipping into the wound, may remain attached to the feather, and the air can escape through a large hole in the upper end.

Prevention is vastly better than cure, and may be always secured, by not allowing hungry cattle to fill themselves with clover, roots, apples, &c. When first put upon such feed, it should be when the dew and rain are off, and their stomachs are already partially filled; and they should then be withdrawn before they have gorged themselves.

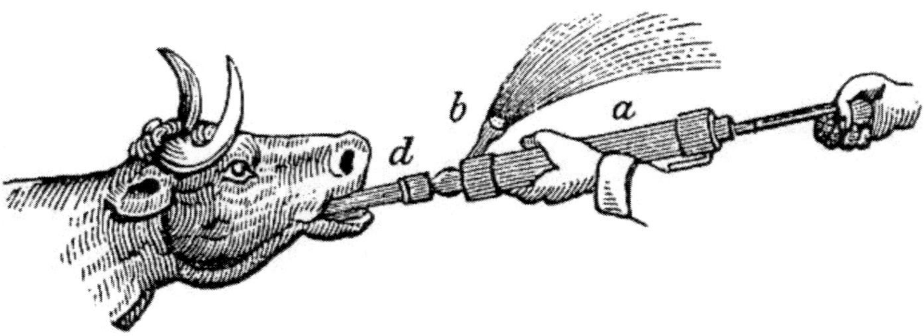

Figure 15.

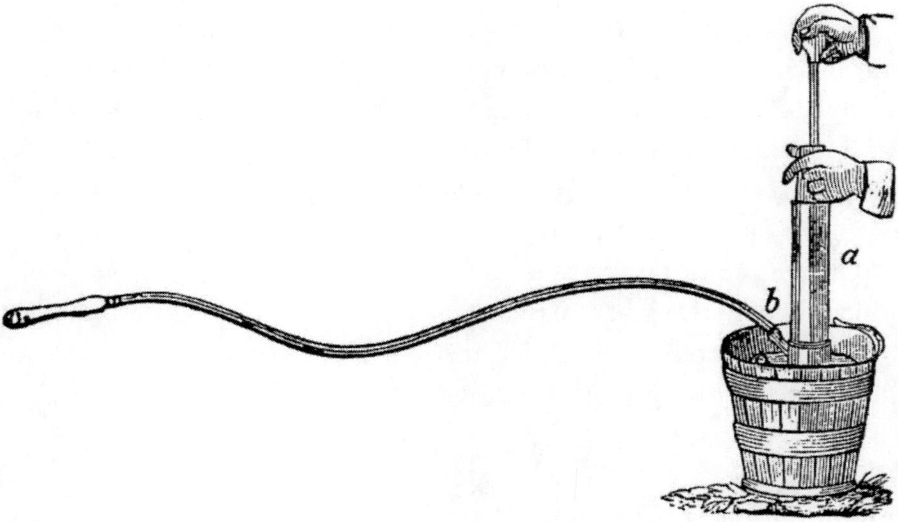

Figure 16. The Stomach Pump.

This is a convenient instrument for extracting poisonous substances from the stomach. It is also highly useful for administering medicines and injections, and if fitted with several tubes, one may suffice for animals of any size. It consists of a syringe, *a*, with a side opening at *b*, and another at the bottom *d*, as shown in Figure 16. For injections, Figure 15 is used, and the end of the syringe is placed in a vessel containing the fluid, when a probang or injection-tube is screwed on to the side opening at *b*, through which the fluid is forced into the stomach or rectum, as may be required. The probang should be a tube of thick but elastic leather, and it may be passed into the mouth, through an aperture in a block, placed on edge between the teeth, which is easily done while a person holds the head of the animal firmly.

Choking

Is frequently relieved by some of the following expedients.

The use of the probang or whip-stock, mentioned under the head of remedies for Hoven, by which the root is forced into the stomach.

A soft root may be crushed so as to allow of swallowing, by holding a smooth block against it, and striking with a mallet on the opposite side.

If within arms-length, the root may be removed by hand.

It is said this can be done, by tying up the fore-leg with a small cord, close to the body, and giving the animal a sudden start with a whip; or by jerking the fore-leg out forward.

Or pour down the throat a pint bottle full of soft soap, mixed with sufficient hot water to make it run freely.

Prevention consists in cutting the roots; not feeding them when the animals are very hungry, and not disturbing them while eating.

Inflammation of the Stomach

This is frequently produced by a sudden change from dry to green food, and some other causes.

Epsom salts, castor oil, sulphur, and carbonate of soda, in sufficient quantity to purge freely, are good remedies.

It may be prevented by changing the food gradually.

Mange, or Scab

This is denoted by the animal rubbing the hair off about the eyes and other parts. The skin is scaly or scabby, sometimes appearing like a large seed-wart.

Remedies.—Rub the spots with sulphur and lard, after scraping and washing with soap.

When the skin is cracked, take sulphur, 1 lb.; turpentine, ¼ lb.; unguentum, (or mercurial ointment,) 2 ounces; linseed oil, 1 pint. Melt the turpentine and warm the oil, and when partly cooled, stir in the sulphur; when cold, add the unguentum, mixing all well. Rub this thoroughly with the hand on the parts affected.

We have no doubt this, like scab in sheep and itch in the human species, will be found, on close investigation, to be caused by minute insects located in the skin. Salt and water ought, in that case, to be a good remedy.

Hollow Horn, or Horn Ail

This is not unfrequently hollow stomach, and very often follows stinted fare, hard usage, and exposure to cold. We have noticed this as most prevalent among oxen that have done a severe winter's work.

Symptoms.—Bloody urine; swollen udder; shaking the head; eyes and head swollen; standing with the head against a fence or barn; eyes dull and sunken, and horns cold.

Remedies.—Bleed and physic, shelter and feed properly.

Take a half pint of good vinegar, two tablespoonfuls of salt, one teaspoonful of pepper, and mix and pour into each ear, holding the head on one side for two minutes.

Bore with a large gimlet on the under side of the horn, three or four inches from the head; and if hollow, bore nearer the head and let out all the matter, and syringe two or three times a day with salt and water, or soap-suds, or salt and vinegar.

Spirits of turpentine rubbed in around the base of the horns, will arrest the disease in its incipient stages.

Pour a spoonful of boiling hot brimstone into the cavity between the horns.

Pour a teakettle of boiling water on the horns, holding so as to prevent injury to the other parts. Soot and pepper given internally are good.

Jaundice, or Yellows

This is owing to gall-stones or calculi, which occasionally accumulate in large numbers, and is sometimes owing to increased or altered quality of the bile. It is manifested by the yellowness of the eye and skin, and high color of the urine, and poor appetite.

Remedies.—Bleed, and purge with Epsom salts.

If taken in season, 2 ounces of ground mustard may be mixed with a liquid, and given twice a day.

Green food is a good preventive.

Mad Itch

This disease exists in some of the Western states, and shows itself by jerking of the head, and itching around the nose and base of the horns. They will lick their sides and backs, and jerk and hiccup till they fill themselves with wind; afterwards they froth at the mouth, and in 24 hours die raving mad.

Remedy.—Give as much soot and salt as the animal will eat; soon after, give ¾ or 1 lb. of brimstone or sulphur; and 8 hours after, as many salts.

Bloody Murrain, or Red Water

This disease first shows itself in a cough, then heaving of the flanks, with bloody, black, and fœtid evacuations, tenderness over the loins, and coldness of the horns. Tumors and biles sometimes appear. The animal holds down the head, moans, is restless, and staggers when walking.

Causes.—We have lost several animals by this fatal disease, and are not aware of having cured any when severely attacked. In repeated instances, we have seen large flukes taken out of the liver, strongly resembling the common leech, which abounds in many of our swampy lands. It is certain that on new, low swamps and clay lands, cattle are most liable to it; and when they have been subject to repeated attacks in such localities, clearing and draining have checked it.

Youatt attributes it to certain kinds of forage, which are peculiar to the above situations. We are rather inclined to ascribe it to exposure, to excessive dampness, and especially to miasma; for although the brute creation are perhaps less sensitive to these influences than man, yet, as they are governed by the same unvarying laws of nature, when subjected to conditions totally unsuited to their economy, they must suffer equally in kind, though probably not in degree, with the more refined human frame. But it is evident the disease, its causes, and remedies, are as yet imperfectly understood.

Remedies.—However intelligent men may differ as to its causes, all agree that the animal should first be bled, and then thoroughly purged. In obstinate cases, this last is a difficult matter. We have given repeated doses of powerful cathartics without producing any effect; and whenever the medicine is inoperative, death speedily follows.

Large doses of common salt, or Epsom salts dissolved in water, are good purgatives, and if the animal neglects drinking after taking them, he should be drenched with copious draughts of water. These should be repeated every few hours, if ineffectual.

Injections are sometimes useful, when medicine fails to act. These may be made of soap and water; or take 2 or 3 gills of oats boiled, 3 drachms saltpetre, 1½ oz. linseed oil, mix and use them when warm.

The opening of the bowels may be followed with a pint of linseed oil, as an additional and gentle laxative.

When the animal begins to recover, gentle astringents and tonics may be given.

Preventives.—We have more confidence in preventives than in remedies. Good keep, shelter, dryness, and clean pastures, will generally prevent attack. The cattle should at all times be supplied with two or three troughs under cover, on the sides and bottoms of which tar should be plentifully spread. Let equal portions of salt and slaked lime be in one; salt and wood ashes in another; and salt and brimstone in a third. Many farmers have entirely avoided this disease while using one or more of these, when they annually lost many by it previously.

Hoof Ail

Is indicated by lameness, fever, and a soft swelling just above the hoof.

Remedies.—Carefully wash the foot in warm soap-suds, and while still damp, apply between the claws on the affected part from one to three grains of corrosive sublimate. If it does not fully adhere, it must be mixed with hog's lard, but it should be so applied as to be out of the reach of the animal's tongue, as it is a powerful poison, and the extreme irritability of the feet will induce him to lick them.

The claw is efficiently cleansed, by drawing a cord briskly through it, when either of the above applications, or blue vitriol put on two or three times a day, or spirits of turpentine, will effect a cure.

It is sometimes cured by putting the animals in the stanchions, and applying a sharp chisel three-fourths of an inch from the toe, and striking it with a mallet till it is cut off. If it does not bleed freely, cut off shavings till it does. If the animal is refractory, let a person hold up the opposite foot. Keep them in the stable two or three days, and out of the mud for a week.

Loss of Cud

Is loss of appetite, prostration, and general ill-health.

Remedies.—Give a warm bran mash, with good hay, and warm water with salt.

An aloe tincture, made with brandy and ginger, is good.

Afterwards give good, dry, nourishing food; and bitter infusions, chamomile flowers, hoarhound, oak bark, &c., in beer.

Scours, or Diarrhœa

A common remedy, is to boil the bark of white oak, white pine, and beech, and give a strong infusion in bran. If they refuse to eat it, pour it down.

The oak is astringent, and the pine and beech soothing and healing.

Warbles

Are grubs, the egg of which is deposited in the back of cattle by the gad-fly, (*Œstrus bovis.*) They are discernible by a protuberance or swelling on the back. They may be pressed out by the thumb and finger; or burnt out by plunging a hot wire in them; or a few applications of strong brine will remove them.

Wounds

In cattle are readily healed, when the animal's blood is in good order, by applying a salve made of 1 oz. green copperas; 2 oz. white vitriol; 2 oz. salt; 2 oz. linseed oil; 8 oz. molasses. Boil over a slow fire 15 minutes in a pint of urine, and when almost cold, add 1 oz. oil of vitriol, and 4 oz. spirits turpentine. Apply it with a feather to the wound, and cure soon follows.

Milk, or Puerperal Fever

Is a common disease with cows in high condition, at the time of calving. It may, in almost every case, be avoided, by keeping them in moderate feed and flesh.

Remedies.—Bleed freely, say 6 to 10 quarts, according to the circulation of the blood; then give 1 to 1½ lbs. of Epsom salts, according to the size of the beast, to be repeated in half lb. doses every six hours, till she purges freely.

Injections should always be given when purgatives are tardy in their operation.

Caked Bag

May be removed by simmering the bark of the root of bitter-sweet in lard, till it becomes very yellow. When cool, apply it to the swollen udder once in 8 or 10 hours; or wash it several times a day in cold water.

A pint of horseradish, fed once a day, cut up with potatoes or meal, is useful for the same purpose.

This is also a tonic, helps the appetite, and is good for oxen subject to heat.

Garget

Is a more intense degree of inflammation than exists in caked bag and sore, swollen teats, and shows itself in hard bunches on the udder.

The cow should be bled, and take a large dose of physic; then wash the udder as in caked bag.

Repeated doses of sulphur is a good remedy.

Garget, or scoke root, given of the size of a large finger, grated and fed in their food, is a general application with farmers. The garget plant grows from three to six feet high, with a purple stalk, and strings of berries hanging down between the branches.

Sore Teats

May be healed by rubbing with goose oil, cream, new milk; or make the same applications for it as for caked bag. The bag and teats should be well cleansed with warm soft water, if to be followed by any ointment.

The following application is recommended by Youatt: One ounce of yellow wax and three of lard; melt together, and when cooling, rub in one quarter ounce of sugar of lead, and a drachm of alum finely powdered.

Warts

Are of two kinds; the first, on the outer skin, may be removed by rubbing with camphorated olive oil. The others penetrate into the flesh, and may be removed by a ligature of fine twine, or silk, or india-rubber drawn into a string, and tied tightly around the wart, which falls off in a few days.

Remedies.—Nitrate of silver, (lunar caustic,) applied to the wart, will remove it, but it produces a sore.

Apply a strong wash of alum.

Rub with the juice of milk-weed.

Poultice with grated carrot.

Cut off the wart with sharp scissors, when the cow is dry. It will bleed little, and soon heal.

Sore Necks on Working Oxen

These occur when worked in wet weather, or with bad yokes. The remedy is, rub with a healing application. The preventive is, good yokes; the application of grease; or a decoction of white or yellow oak bark applied to the affected parts. Or, a better preventive is a canvass or leather cap to protect the neck entirely from the storm.

The Bite of Poisonous Snakes

May be cured by shaking together equal parts of olive oil and hartshorn, and rubbing the wound and adjacent parts three or four times a day. For a full-grown animal, one quart of olive oil and an ounce of hartshorn should be administered internally, in addition to the above.

For Stings of Bees, Hornets, &c.

Apply warm vinegar and salt, rubbing the parts thoroughly.

For a Forming Tumor

Rub thoroughly with strong brine, or a solution of sal ammoniac dissolved in eight times its weight of water. If the tumor comes to a head, open it near the bottom with a lancet; or place a seton in it so as to admit the escape of purulent matter.

Lice and Vermin

Sometimes abound on cattle during the latter part of winter and spring. These are generally the result of *mange*, which is itself the effect of ill-feeding and ill-condition. They are removed with the cause. We doubt if they can be permanently kept off, where the animal is losing flesh and health.

Remedies.—Restore the health and condition, and sprinkle sand, ashes, or dirt plentifully around the roots of the horns, and along the ridge of the neck and back.

A liberal application of train or other oil has nearly a similar effect.

Never apply an ointment containing corrosive sublimate or other poison, as it may be licked by the animal or its fellows, who may thus become seriously poisoned.

The Trembles

Producing milk sickness (a most fatal disease) in the human family, from eating the milk or flesh of animals affected by it. This disease, which exists principally in the region of the Wabash River, is supposed by Dr.

Drake to be owing to the poison oak, (Rhus Toxicodendron,) or poison vine, (Radicans,) which the animals eat.

Symptoms.—The animal mopes, is feverish and costive, but apparently preserves its appetite. The next stage of the disease is faintness and vertigo, which is shown when the animal is put upon exertion, being followed by excessive trembling and entire prostration.

Remedy.—Almost every cathartic has been tried in vain. Indian corn, both dry and green, has been fed to all animals accustomed to eating it, and when they can be induced to feed upon it freely, purging is generally secured. Rest of the animal is absolutely essential while the disease continues, and is itself an effectual remedy in mild cases.

Besides the diseases enumerated, there are occasional epidemics, such as black tongue, black foot, or foot root, &c., which carry off great numbers of animals. Remedies for these are frequently not discovered, and the epidemic is allowed to run its course unchecked. The only preventives are such care, food, and management as the experienced herdsman knows to be best suited to the maintenance of the health and thrift of his stock.

Note.—Some ailments will be found under the head of diseases of the other animals mentioned in this work, the general resemblance of which to each other will justify nearly a similar treatment.

If intelligent farriers are at hand, they may sometimes be called in with advantage; though we acknowledge our distrust of the quackery of most of those passing under this title. There is little science or intelligent study in the composition of this class, the world over; and much of their practice is the merest empiricism. The owner should see to it, if he employs one of whose attainments he is doubtful, that neither medicines nor operations be used, unnecessarily severe or hazardous to the animal. Especially, should the diabolical practice be interdicted, of the abundant and indiscriminate use of poisons, boiling oils, turpentine, and tar, and the hot iron applied to the sensitive wound or naked flesh. If certain or effectual remedies for the removal of disease cannot he applied, such as augment the suffering or endanger the life of the poor dumb things, may at least be avoided.

Chapter 3

The Dairy

Cows for the Dairy

From what has been said on the various characteristics of the different breeds of cattle, it must be evident, that no very definite criteria of excellence can be given for all good dairy cows. But there are certain points in a good milker, that can hardly be mistaken.

She should be descended from the best milking stock; her head should be small or of medium size, muzzle fine, and nostrils flexible and expanded; face long, slender, and dishing; cheeks thin; eyes full, mild, and prominent; horns delicate and waxy, and they may be either branching, lopped, crumpled, or hornless; long, thin, lively ear, and the inside of an orange color; neck thin and small at its junction with the head; deep chest, but not too heavy before; back level and broad; well ribbed; belly large; low flank; wide thighs, but thin; short legs, and standing well apart; large milking veins; loose, capacious udder, coming well out behind; good teats; loose, mellow skin, of a deep yellow; and a fine, thick coat of glossy hair; and she must be of a good disposition, and free from tricks.

Yet, with all the skill of a well-practised taste in the selection of animals, the dairyman will frequently find his theories and results at sad variance. One may sometimes select a fine animal, with every appearance of good milking qualities, which is but a medium cow at the pail; and another, that hardly seems worthy of notice, and which sets at defiance many established milking points, and all preconceived notions of symmetry, may yet prove a good milker. A cow that runs to flesh while in milk, is generally an indifferent animal for the dairy. Perfection in a cow, consists in converting all she eats into milk while yielding it, and when dry, in turning all she consumes into valuable meat.

Management of Dairy Cows

A cow may have her first calf when between two and three years of age, according to her size and development. After calving, she should be stinted in her food for two or three days, and not fed freely for a week. Avoid fat in a breeding cow. Too high feeding is the cause of milk-fever, caked bag, garget, and a host of evils; and very poor feed is almost equally objectionable. The average time of a cow with young, is from 40 to 41 weeks; but they sometimes go only 34, and occasionally overrun 44.

A dry, unoccupied stall or yard is best for her to calve in; and if there is any serious delay or difficulty in the operation, she may be assisted by placing the fœtus in the right position, and gently pulling it, with every throe of the dam. After the calf has drawn all he wants at morning and evening, the bag should be thoroughly and quickly emptied of all the milk. If strong and vigorous, the calf is the best doctor for garget or caked bag. He may be allowed to suck the cow or not, at the option of the owner; there are reasons for and against the practice, as will be seen under the head of rearing calves, and each person must determine in his own case on which side the balance lies.

Milking

This is an important operation, and on its proper performance depends much of the success of the dairyman. A cow regularly, gently, yet quickly and thoroughly milked, will give much more than if neglected. If a herd of cows be separated into two divisions, each yielding the same quantity of milk, and one is given to a good milker, and the other to a shiftless or lazy one, the latter will speedily reduce his milk much below the quantity obtained by the former; and if the milkers then exchange cows, they will be found to change quantity too, those before affording the least, soon giving the most. An indifferent milker ought never to be tolerated in a herd, good ones are cheaper at double the price. It is best to milk at intervals of about 12 hours; which may be done when pastures are convenient, or cows are soiled or fed in the yard. But as this is not often the case in the season of green food, they should be milked early in the morning and turned into pasture, to fill themselves before the sun is oppressive; and if they are to be kept up at night, let them browse in the pasture as long as possible, before they are brought to the yard.

MILK

Is produced from the females of all the warm-blooded animals, which are enumerated among the mammaliæ. The milk of several animals is employed for domestic purposes, among different nations. That of the camel is used by the Arabs; the milk of the ass by the Spaniards, the Maltese, and the inhabitants of the Levant; that of the mare by the Cossacks, the Kirgheez, and other Tartars; and that of the goat, the ewe, and the cow, by most of the ancient, and with few exceptions, by every modern European nation. Within the last century, however, the use of all excepting cow's milk has been almost entirely discarded, among the most highly civilized people.

If we except some few Welsh and Swiss, or other emigrants, who resort to the goat and ewe for their dairy materials, for the first few years of their residence here, the cow is the only animal which is employed in America for producing milk. For this, she is pre-eminently fitted, and the modern improvement of this invaluable animal has carried her product of milk almost as far as can be reasonably looked for from a given amount of food; and although this is of about the average richness of the goat and ewe, and before that of the ass, the quantity she yields is frequently as 60 to 1, in favor of the cow, over the first two competitors. As a milk-giving animal, the cow is the best fitted for the purposes of civilized man; and she is made to contribute, not only to his health, his comfort, and his economy, but to many of his choicest luxuries. Milk contains every element of nutrition necessary to animal existence; and man can subsist, with unimpaired health and strength, if limited to this food alone.

The Constituents of Milk

Are butter, which varies from 2 to 6 per cent.; casein or cheese, usually 4 to 5, but sometimes varying from 3 to 15 per cent.; (the last excessive quantity, yielded only by the first milk after calving;) milk-sugar, 4 to 6; salts or saline matter, 0.2 to 0.6; and water, 80 to 89.

There is much diversity in the product and quality of milk from cows of the same breed, the same food, and other circumstances and conditions, apparently equal. Thus, of a herd of 22, chiefly Ayrshire, one gave 84 quarts in one week, which afforded 3½ lbs. of butter; two others in the same time gave 86, yielding 5½ lbs.; and a fourth gave 88 quarts, making 7 lbs.

The amount of butter, however, which a given quantity of milk will produce, is not the only criterion of the value of the milk, except for this purpose alone. Some cows will yield more butter, others will produce more cheese; while for consumption, another may partially compensate, in the increased quantity of milk-sugar, and the saline matters, for a deficiency of both the other ingredients. But for dairy purposes, butter and cheese are the only measure of the value of milk; and a cow is esteemed good or indifferent, as she gives one or the other in the greatest abundance.

Circumstances Which Modify the Quantity and Character of Milk

Besides the accidental variation in the quantity and quality of milk in different animals, before adverted to, there are many reliable causes which influence both. Of these, parentage has a most decided and uniform influence, frequently modified, however, in the particular individual, by some personal and controlling causes. But a cow, whose maternal ancestry on both sides are choice milkers, is almost certain to resemble them.

Food influences the quantity, rather than the quality. Boussingault tried numerous experiments, with cows fed on various kinds of food, and found the difference hardly appreciable in the quality of milk. Its true benefit is to be looked for in the increased quantity, through which, the valuable ingredients are distributed in nearly the same proportion, as when the product is materially lessened. By quality we mean to be understood, the amount of the ingredients, valuable for nutrition only; for it is certain, that there is a rich aromatic flavor, not only in milk, but in butter and cheese, which is afforded in various articles of food, and especially by the fresh green herbage which abounds in the pastures from spring to autumn.

Activity or rest has a great effect on both quantity and quality. The less action, and the more quiet and rest, the greater the amount of milk and butter. But exercise is absolutely essential to the production of cheese. Butter may be made from cows confined in a stable, but cheese can only be profitably made from animals at pasture. It is supposed by physiologists, that the exercise in gathering their food, rather than any peculiarity in its character, is necessary to convert the nitrogenized tissues into the nitrogenized principle of caseum or cheese.

The time from calving, has also its effect. The first milk drawn from a cow after calving, has been found to yield over 15 per cent. of casein, while in its ordinary state it gives only three to five and a half. As the quantity of milk diminishes in a farrow cow, the quality improves within certain limits. Pregnancy affects the quality injuriously, and especially towards its latter stages; and a cow that is predisposed to giving milk, should be dried off a few weeks before its expiration, as it is then unfit for use.

Fat cows give poorer milk than such as are moderately lean; and young animals do not come up to the maximum of their quality, till after their third or fourth calving.

The milk first drawn from the udder, will yield only an eighth, and sometimes even a much less proportion of cream, than the strippings; and the milk which is drawn three times a day, is greatly inferior to such as is taken but once, though the latter is less abundant.

Excitement, or fretfulness; change of locality, or to a different herd, with new companions; separation from her calf; periodical heat; annoyance from flies, or worry from dogs; exposure to storms, severe cold, or an oppressive sun, and many similar causes, diminish the quantity of milk and butter; but some of these may reasonably be expected to increase the proportion of its casein.

Dr. Playfair found that the quantity of butter in the evening milk, after the cow had been at pasture all day, was 3.7 per cent., while the casein was 5.4; after lying quietly all night, the milk from the same cow, on the following morning, contained 5.6 per cent of butter, and only 3.9 of casein. In stabling the cow, the butter was invariably in greater proportion than when allowed to ramble in the pasture; and the casein, with a single exception, was equally diminished.

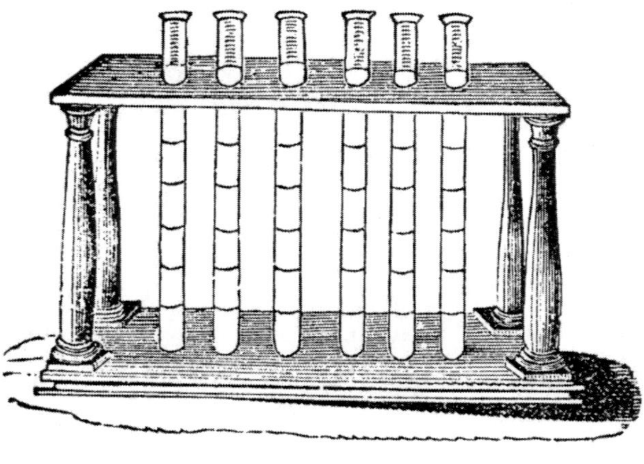

Figure 17. Lactometer, or Cream Guage.

Figure 17, is a number of glass tubes of equal size, set in a frame called a lactometer or cream guage. If milk from different cows be set in these, the depth of the cream will indicate their comparative richness.

Cream

If milk be immediately set away in shallow vessels, after being taken from the cow, the cream rises to the surface, carrying with it most of the butter contained in the milk, and much of its casein also. Hence, the great nutritive properties of buttermilk, which retains the casein in very large proportions, much of it being rejected by the butter in its separation from the cream.

A temperature below 34°, will prevent the cream from rising in any considerable quantity, and preserve the milk unaltered for some weeks. Coagulating the milk from any cause, will equally prevent the separation of the cream. The elevation of temperature within certain limits, hastens the separation. Thus, at 50°, the cream will mostly have risen in 36 hours; at 55°, in 24; at 68°, in 18 or 20; and at 77°, in 10 or 12 hours.

Heating the milk near the boiling point, and then setting it away and allowing it to remain undisturbed, will soon cause the cream to rise. In the celebrated Orange dairy, near Baltimore, Md., this system was practised, by which, not only most of the cream was secured for butter, but in consequence of its rapid separation, the skimmed milk was sent to market within a few hours after being drawn; and the scalding imparted to it an agreeable flavor and apparent richness, which it did not really possess.

The celebrated clouted cream of Devonshire, England, and the butter made from it, contains an unusual quantity of casein, the consequence of heating the milk. "It is prepared by straining the warm milk into large shallow pans into which a little water has previously been put, allowing these to stand from six to twelve hours, and then carefully heating them over a slow fire, or on a hot plate, till the milk approaches the boiling point. The milk, however, must not actually boil, nor must the skin of the cream be broken. The dishes are now removed into the dairy, and allowed to cool. In summer the cream should be churned on the following day; in winter it may stand over two days. The quantity of cream obtained is said to be one-fourth greater by this method, and the milk which is left is proportionably poor."— [*Johnston.*]

BUTTER

Electricity

Has much to do with the changes in milk, as in all other substances. Glass milk-pans might, therefore, be supposed to be the best vessels for keeping the milk unchanged and sweet.

It may possibly not afford any practical result, yet intelligent experiments for introducing a stream of electricity might well be justified, to aid in the separation of the butteraceous particles in the operation of churning.

Sour Cream

Cream, for the purpose of churning, is usually allowed to become sour. It ought to be at least one day old, but may with advantage be kept several days in cool weather, if it be previously well freed from milk, and be frequently stirred to keep it from curdling. This sour cream is put into the churn, and worked in the usual way until the butter separates. This is collected into lumps, well beat and squeezed free from the milk, and in some dairies is washed with pure cold water as long as the water is rendered milky. In other localities the butter is not washed, but after being well beat, is carefully freed from the remaining milk by repeated squeezings and dryings with a clean cloth. Both methods, no doubt, have their advantages. In the same circumstances, the washed butter may be more easily preserved in the fresh state, while the unwashed butter will probably possess a higher flavor.

Sweet Cream

May be put into the churn and the butter be obtained, but in most cases it requires more labor and longer time, without, in the opinion of good judges, affording in general a finer quality of butter. In all cases the cream becomes sour during the agitation, and before the butter begins distinctly to form.

Clouted Cream

The churning of the clouted cream of this and other countries, forms an exception to the general rule just stated, that more time is required in the churning of sweet creams. Clouted cream may be churned in the morning after it is made, that is, within twenty-four hours of the time when the milk was taken from the cow; and from such cream it is well known that the butter separates with very great ease. But in this case, the heating of the cream has already disposed the oily matter to cohere, an incipient running together of the globules has probably taken place before the cream is removed from the milk, and hence the comparative ease with which the churning is effected. There is something peculiar in butter prepared in this way, as it is known in other countries by the name of Bohemian butter. It is said to be very agreeable in flavor, but it must contain more cheesy matter than the butter from ordinary cream.

Churning the Whole Milk

Is a much more laborious method, from the difficulty of keeping in motion such large quantities of fluid. It has the advantage, however, of giving a larger quantity of butter. At Rennes, in Brittany, the milk of the previous evening is poured *into the churn* along with the warm morning's milk, and the mixture is allowed to stand for some hours, when the whole is churned. In this way it is said that a larger quantity of butter is obtained, and of a more delicate flavor.

In the neighborhood of Glasgow, according to Mr. Aiton, the milk is allowed to stand six, twelve, or twenty-four hours in the dairy, till the whole has cooled, and the cream has risen to the surface. Two or three milkings, still sweet, are then poured together with their cream, into a large vessel, and are left undisturbed till the whole has become quite sour, and is completely coagulated. The proper sourness is indicated by the formation of a stiff *brat* upon the surface *which has become uneven*. Great care must be taken to keep the brat and curd unbroken until the milk is about to be churned, for if any of the whey be separated, the air gains admission to it and to the curd, and fermentation is induced. By this fermentation, the quality of the butter may or may not be affected, but that of the buttermilk is almost sure to be injured.

In Holland the practice is a little different. The cream is not allowed to rise to the surface at all, but the milk is stirred two or three times a day, till it gets sour, and so thick that a wooden spoon will stand in it. It is then put into the churn, and the working, or the separation of the butter is assisted by the addition of a quantity of cold water. By churning the sour milk in one or other of these ways, the butter is said to be "rich, sound, and well-flavored." If it be greater in quantity, it is, according to Sprengel, because the fatty matter carries with it from the milk a larger quantity of casein than it does in most cases from the cream alone.

Sourness of the Cream

For the production of the best butter, it is necessary that the cream should be sufficiently sour before it is put into the churn. Butter made from sweet cream (not clouted) is neither good in quality, nor large in quantity, and longer time is required in churning. It is an unprofitable method.

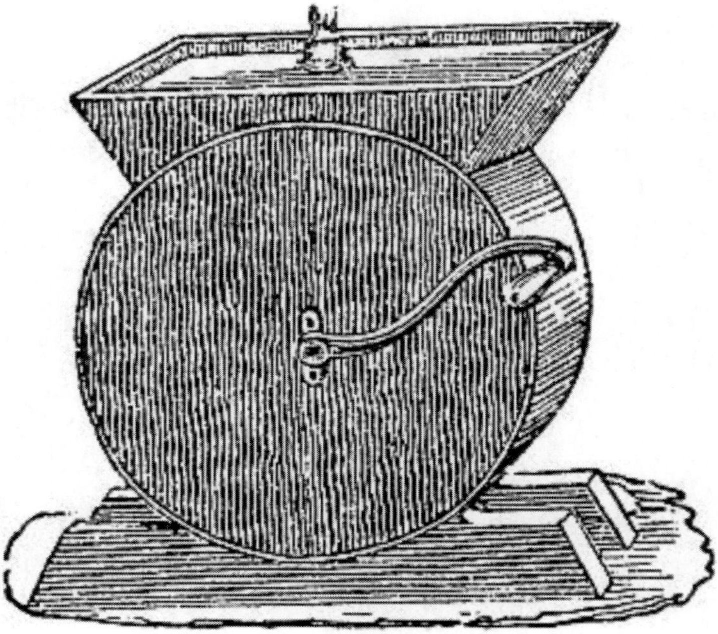

Figure 18.

Figure 18 is a Cylindrical Thermometer Churn, of any required size, with false metal bottom to hold cold or hot water for bringing the cream to the proper temperature. A thermometer, permanently set in the side, indicates the heat.

Quickness in Churning

The more quickly milk or cream is churned, the paler, the softer, and the less rich the butter. Cream, according to Mr. Aiton, may be safely churned in an hour and a half, while milk ought to obtain from two to three hours. The churning ought always to be regular, slower in warm weather, that the butter may not be soft and white, and quicker in winter, that the proper temperature may be kept up.

A barrel-churn, lately introduced into this country, being placed in a trough of water of the proper temperature, readily imparts the degree of heat required by the milk or cream without the necessity of adding warm water to the milk, *and churns the whole in ten or twelve minutes*. It is said also to give a larger weight of butter from the same quantity of milk. If the quality be really as good by this quick churning, the alleged inferiority in the quality of butter churned quickly in the common churn cannot be due to the mere rapidity of churning alone.

Over-Churning

When the process of churning is continued after the full separation of the butter, it loses its fine yellowish, waxy appearance, and becomes soft and light-colored. The weight of the butter, however, is considerably increased; and hence, in Lancashire, over-churning is frequently practised in the manufacture of fresh butter for immediate sale.

Temperature of the Milk or Cream

Much also depends upon the temperature of the milk or cream when the churning is commenced. Cream when put into the churn should never be warmer than 55° Fahrenheit It rises during the churning from 4° to 10° F. above its original temperature. When the whole milk is churned, the temperature should be raised to 65° F., which is best done by pouring in hot water into the churn *while the milk is kept in motion*. In winter, either of these temperatures may be easily attained. In cold weather it is often necessary to add hot water to the cream to raise it even to 55°. But in summer, and especially in hot weather, it is difficult, even in cool and well-ordered dairies, (without the use of ice,) to keep the cream down to this comparatively low temperature. Hence, if the cream be then churned, a second-rate butter, at best, is all that can be obtained.

The Alleged Advantages of Churning the Entire Milk

The proper temperature can be readily obtained both in winter and summer. A hundred gallons of entire milk, will give, in summer, five per cent more butter than the cream from the same quantity of milk. Butter of the best quality can be obtained without difficulty, both in winter and summer. No special attention to circumstances, or change of method, is at any time required. The churning in winter and summer is alike simple and easy. The butter is not only of the best quality while fresh, but is also best for long-keeping, when properly cured or salted.

Cleanliness in All the Operations of the Dairy

This is peculiarly necessary to the manufacture of good butter. Cream is remarkable for the rapidity with which it absorbs and becomes tainted by any unpleasant odors. It is very necessary that the air of the dairy should be sweet, that it should be often renewed, and that it should be open in no direction from which bad odors can come. (*Johnston and other authorities.*)

The statement of J. T. Lansing, who received the first premium for butter from the New York State Agricultural Society, is as follows:—

> *Keep the cows stabled through the inclement season*; feed them from three to four times per day with good hay or green stalks; when near coming in, add some oats, barley, or corn cracked. In summer, good pasture, with living water accessible at all times, and plenty of salt.
>
> *Treatment of milk and cream before churning.*—Strain the milk in tin pans; place them in a cool cellar for the cream to rise. When sufficiently risen, separate the cream from the milk; put in stone jars, well prepared before churning.
>
> *The mode of churning in summer.*—Rinse the churn with cold water; then turn in the cream, and add to each jar of cream put in the churn, full one-fourth of the same quantity of cold water. The churn used is a patent one, moved by hand with a crank, having paddles attached, and so

constructed as to warm the milk (if too cold) with hot water, without mixing them together. The milk and cream receive the same treatment in winter as in summer; and in churning, use hot instead of cold water, if necessary.

The method of freeing the butter from the milk, is to wash the butter with cold water, till it shows no color of the milk, by the use of a ladle.

Salting the butter.—Use the best kind of Liverpool sack-salt; the quantity varies according to the state in which the butter is taken from the churn; if soft, more; if hard, less; always taking the taste for the surest guide. Add no saltpetre, nor other substances.

The best time for churning is the morning, in hot weather, and to keep the butter cool till put down.

The best mode of preserving butter, in and through the summer and winter, is as follows:—The vessel is a stone jar, clean and sweet. The mode of putting it down is to put in a churning of butter, and put on strong brine; let it remain on until the next churning is ready to put down, and so on till the jar is filled; then cover it with fine salt the same to remain on till used.

Mr. McWilliams, of Orange county, the celebrity of whose butter is unsurpassed, thus details his method of butter-making:

"Our practice is not to churn the milk until it becomes thick or loppered, the milk and cream is then churned together. The temperature of the milk is about fifty degrees. In warm weather about a quart of cold water is put in each pan before the milk is strained, so as to keep it sweet as long as possible. The cellar-floor is brick. This in warm weather is daily cleansed with cold water. A drain from the cellar carries off the water thus applied. The churn is filled about half full with milk, with the addition of two pails of cold water before starting the churn. In cold weather the same quantity of warm water is applied. When the churning is finished, which usually occupies about two hours of time, there are then two more pails of cold water applied to raise the butter and cool it.

The butter is then taken out of the churn and put in a large tray; this is immediately filled with cold water, and the butter carefully washed; after which the water is thrown off. The butter now undergoes the process of salting; it is then placed in a cool situation, where it stands about an hour, and is worked carefully over. This finished, it is placed in the same situation as before, where it stands three or four hours, and is again worked over; again replaced for five or six hours, when it is worked over for the third time. It is now replaced, where it stands till the next morning, and worked over for the fourth time. A small quantity of nitre is then put in the butter. Thus finished, it is placed in firkins holding about eighty-five pounds.

Previous to packing, the firkin is scalded with hot water, rinsed and cooled with cold water, then rubbed all around with fine salt; this prevents the butter from adhering to the sides of the firkin. When the firkin is full, a linen cloth is placed over the top of the butter; on this cloth a covering of salt is put one inch deep, and cold water enough added to it to form a brine. It then stands till it is to be sent to market, when the cloth and salt are removed, the firkin turned down, the top of the butter in the keg washed with cold water, and the pickle drained off. The firkin is now neatly headed up and sent to market."

The salt added to the butter should be from 1-24th to 1-28th of its weight, or about two-thirds of an ounce to a pound, and this must be of the best quality. All the buttermilk must be thoroughly extracted by repeated washings; and when completed, the butter should be immediately packed, and not a particle of air allowed to come in contact with it till opened for the table.

CHEESE

The Circumstances Affecting the Quality of Cheese

"All cheese consists essentially of the curd, mixed with a certain portion of the fatty matter, and of the sugar of milk. But differences in the quality of the milk, in the proportion in which the several constituents of milk are mixed together, or in the general mode of dairy management, give rise to varieties of cheese almost without number. Nearly every dairy district produces one or more qualities of cheese peculiar to itself.

Natural Differences in the Milk

It is obvious that whatever gives rise to natural differences in the quality of the milk, must affect also that of the cheese prepared from it. If the milk be poor in butter, so must the cheese be. If the pasture be such as to give a milk rich in cream, the cheese will partake of the same quality. If the herbage or other food affect the taste of the milk or cream, it will also modify the flavor of the cheese.

Milk of Different Animals

So the milk of different animals will give cheese of unlike qualities. The ewe-milk cheeses of Tuscany, Naples, and Languedoc, and those of goats' milk made on Mont Dor and elsewhere, are celebrated for qualities which are not possessed by cheeses prepared from cows' milk in a similar way. Buffalo milk also gives a cheese of peculiar qualities, which is manufactured in some parts of the Neapolitan territory."

Other kinds of cheese are made from mixtures of the milk of different animals. Thus the strong-tasted cheese of Lecca and the celebrated Roquefort cheese are prepared from mixtures of goat with ewe milk, and the cheese of Mont Cenis from both of these mixed with the milk of the cow.

Creamed or Uncreamed Milk

Still further differences are produced, according to the proportion of cream which is left in or is added to the milk. Thus, if cream only be employed, we have the rich *cream-cheese* which must be eaten in a comparatively recent state. Or, if the cream of the previous night's milking be added to the new milk of the morning, we may have such cheese as the *Stilton* of England, or the small, soft, and rich *Brie* cheeses, so much esteemed in France.

If the entire milk only be used, we have such cheeses as the *Cheshire*, the *Double Gloucester*, the *Cheddar*, the *Wiltshire*, and the *Dunlop* cheeses of Britain, the Kinnegad cheese of Ireland, and the Gouda and Edam cheeses of Holland. Even here, however, it makes a difference, whether the warm milk from the cow is curdled alone, as at Gouda and Edam, or whether it is mixed with the milk of the evening before, as is generally done in Cheshire and Ayrshire. Many persons are of opinion that cream, which has once been separated, can never be so well mixed again with the milk, that a portion of the fatty matter shall not flow out with the whey and render the cheese less rich.

If the cream of the evening's milk be removed, and the skimmed milk added to the new milk of the next morning, such cheeses as the *Single Gloucester* are obtained. If the cream be taken once from *all* the milk, the better kinds of skimmed-milk cheese, such as the Dutch cheese of Leyden, are prepared; while if the milk be twice skimmed, we have the poorer cheeses of Friesland and Groningen. If skimmed for three or four days in succession, we get the hard and horny cheeses of Essex and Sussex, which often require the axe to break them up.

Buttermilk Cheese

But poor or butterless cheese will also differ in quality according to the state of the milk from which it is extracted. If the new milk be allowed to stand to throw up its cream, and this be then removed in the usual way, the ordinary skimmed-milk cheese will be obtained by adding rennet to the milk. But if, instead of skimming, we allow the milk to stand till it begins to sour, and then remove the butter by churning the whole, we obtain the milk in a sour state, (*buttermilk.*) From this milk the curd separates naturally by gentle heating. But being thus prepared from sour milk, and without the use of rennet, buttermilk cheese differs more or less in quality from that which is made from sweet skimmed-milk. The acid in the buttermilk, especially after it has stood a day or two, is capable of coagulating new milk also; and thus, by mixing more or less sweet milk with the buttermilk before it is warmed, several other qualities of mixed butter and sweet-milk cheese may readily be manufactured.

Whey Cheese

The whey which separates from the curd, and especially the white whey, which is pressed out towards the last, contains a portion of curd, and not unfrequently a considerable quantity of butter also. When the whey is heated, the curd and butter rise to the surface, and are readily skimmed off. This curd alone will often yield a cheese of excellent quality, and so rich in butter, that a very good imitation of Stilton cheese may sometimes be made with alternate layers of new-milk curd and this curd of whey.

Mixtures of Vegetable Substances with the Milk

New varieties of cheese are formed by mixing vegetable substances with the curd. A green decoction of two parts of sage leaves, one of marigold, and a little parsley, gives its color to the *green cheese* of Wiltshire; some even mix up the entire leaves with the curd. The celebrated Schabzieger cheese of Switzerland, is made by crushing the skim-milk cheese after it is several months old to fine powder in a mill, mixing it then with one-tenth of its weight of fine salt, and one-twentieth of the powdered leaves of the mellilot trefoil, (*trifolium melilotus cerulea*,) and afterwards with oil or butter, working the whole into a paste, which is pressed and carefully dried.

Potato Cheeses

As they are called, are made in various ways. One pound of sour milk is mixed with five pounds of boiled potatoes and a little salt, and the whole is beat into a pulp, which, after standing five or six days, is worked up again, and then dried in the usual way. Others mix three parts of dried boiled potatoes with two of fresh curd, or equal weights, or more curd than potato, according to the quality required. Such cheeses are made in Thuringia, in Saxony, and in other parts of Germany. In Savoy, an excellent cheese is made by mixing one of the pulp of potatoes with three of ewe-milk curd; and in Westphalia, a potato cheese is made with skimmed milk.

Preparation of Rennet

Rennet is prepared from the salted stomach or intestines of the sucking calf, the unweaned lamb, the young kid, or even the young pig. In general, however, the stomach of the calf is preferred, and there are various ways of curing and preserving it.

The stomach of the newly killed animal contains a quantity of curd derived from the milk on which it has been fed. In most districts, it is usual to remove by a gentle washing the curd and slimy matters which are present in the stomach, as they are supposed to impart a strong taste to the cheese. In Cheshire, the curd is frequently salted separately for immediate use. In Ayrshire and Limburg, on the other hand, the curd is always left in the stomach and salted along with it. Some even give the calf a copious draught of milk shortly before it is killed, in order that the stomach may contain a larger quantity of the valuable curd.

Salting the Stomach

In the mode of salting the stomach, similar differences prevail. Some merely put a few handfuls of salt into and around it, then roll it together, and hang it near the chimney to dry. Others salt it in a pickle for a few days, and then hang it up to dry; while others pack several of them in layers, with much salt both within and without, and preserve them in a cool place, till the cheese-making season of the following year. They are then taken out, drained from the brine, spread upon a table, sprinkled with salt which is rolled in with a wooden roller, and then hung up to dry. In some foreign countries, the recent stomach is minced very fine, mixed with salt and bread into a paste, put into a bladder, and then dried. In Lombardy, the stomach, after being salted and dried, is minced and mixed up with salt, pepper, and a little whey or water into a paste, which is preserved for use.

In whatever way the stomach or intestine of the calf is prepared and preserved, the almost universal opinion seems to be, that it should be kept for 10 or 12 months, before it is capable of yielding the best and strongest rennet. If newer than 12 months, the rennet is thought to make the cheese heave or swell, and become full of eyes or holes.

Making the Rennet

In making the rennet, different customs also prevail. In some districts, a bit of the dried stomach is put into half a pint of lukewarm water, with as much salt as will lie upon a shilling, is allowed to stand overnight, and in the morning the infusion is poured into the milk. For a cheese of 60 lbs. weight, a piece of the size of a dollar will often be sufficient, though of some skins as much as 10 square inches are required to produce the same effect. It is, however, more common to take the entire stomach, and to pour upon them from one to three quarts of water for each stomach, and to allow them to infuse for several days. If only one has been infused, and the rennet is intended for immediate use, the infusion requires only to be skimmed and strained. But if several be infused, or as many as have been provided for the whole season, about two quarts of water are taken for each, and, after standing not more than two days, the infusion is poured off, and is completely saturated with salt. During the summer it is constantly skimmed, and fresh salt added from time to time. Or a strong brine may at once be poured upon the skins, and the infusion, when the skins are taken out, may be kept for a length of time. Some even recommend, that the liquid rennet should not be used until it is at least two months old. When thus kept, however, it is indispensable that the water should be fully saturated with salt.

In Ayrshire, and in some other countries, it is customary to cut the dried stomach into small pieces, and to put it, with a handful or two of salt and one or two quarts of water, into a jar, to allow it to stand for two or three days, afterwards to pour upon it another pint for a couple of days, to mix the two decoctions, and when strained, to bottle the whole for future use. In this state it may be kept for many months.

In making rennet, some use pure water only, others prefer clear whey, others a decoction of leaves, such as those of the sweetbrier, the dog-rose, and the bramble, or of aromatic herbs and flowers; while others again, put in lemons, cloves, mace, or brandy. These various practices are adopted for the purpose of making the rennet keep better, of lessening its unpleasant smell, of preventing any unpleasant taste it might give to the curd, or finally of directly improving the flavor of the cheese. The acidity of the lemon will, no doubt, increase also the coagulating power of any rennet to which it may be added.

The rennet thus prepared is poured into the milk previously raised to the temperature of 90° or 95° F., and is intimately mixed with it. The quantity which it is necessary to add varies with the quality of the rennet, from a tablespoonful to half a pint for 30 or 40 gallons of milk. The time necessary for the complete fixing of the curd varies also from 15 minutes to an hour or even an hour and a half. The chief causes of this variation, are the temperature of the milk, and the quality and quantity of the rennet employed.

Different Qualities of Cheese

The temperature of new or entire milk, when the rennet is added, should be raised to about 95° F.; that of skimmed milk need not be quite so high. If the milk be warmer the curd is hard and tough, if colder, it is soft and difficult to obtain free from the whey. When the former happens to be the case, a portion of the first whey that separates may be taken out into another vessel, allowed to cool, and then poured in again. If it prove to have been too cold, hot milk or water may be added to it; or a vessel containing hot water may be put into it before the curdling commences; or the first portion of whey that separates may be heated and poured again upon the curd. The quality of the cheese, however, will always be more or less affected, when it happens to be necessary to adopt any of these remedies. To make the best cheese, the true temperature should always be attained as nearly as possible, before the rennet is added.

Mode in Which the Milk Is Warmed

If, as is the case in some dairies, the milk be warmed in an iron pot upon the naked fire, great care must be taken that it is not singed or *fire-fanged.* A very slight inattention may cause this to be the case, and the taste of the cheese is sure to be more or less affected by it. In Cheshire, the milk is put into a large tin pail, which is plunged into a boiler of hot water, and frequently stirred till it is raised to the proper temperature. In large dairy establishments, however, the safest method is to have a pot with a double bottom, consisting of one pot within another, after the manner of a glue-pot; the space between the two being filled with water. The fire applied beneath, thus acts only upon the water, and can never, by any ordinary neglect, do injury to the milk. It is desirable in this heating, not to raise the temperature higher than is necessary, as a great heat is apt to give an oiliness to the fatty matter of the milk.

The Time During Which the Curd Stands

This is also of importance. It should be broken up as soon as the milk is fully coagulated. The longer it stands after this, the harder and tougher it will become.

The Quality of the Rennet

This is of much importance, not only in regard to the certainty of the coagulation, but also to the flavor of the cheese. In some parts of Cheshire, it is usual to take a piece of the dried membrane and steep it overnight with a little salt for the ensuing morning's milk. It is thus sure to be fresh and sweet, if the dried maw be in good preservation. But where it is customary to steep several skins at a time, and to bottle the rennet for after-use, it is very necessary to saturate the solution completely with salt, and to season it with spices, in order that it may be preserved in a sweet and wholesome state.

The Quantity of Rennet Added

This ought to be regulated as carefully as the temperature of the milk. Too much renders the curd tough; too little causes the loss of much time, and may permit a larger portion of the butter to separate itself from the curd. It is to be expected also, that when rennet is used in great excess, a portion of it will remain in the curd, and will naturally affect the kind and rapidity of the changes it afterwards undergoes. Thus, it is said to cause the cheese to heave or swell out from fermentation. It is probable, also, that it will affect the flavor which the cheese acquires by keeping. Thus it may be, that the agreeable or unpleasant taste of the cheeses of certain districts or dairies may be less due to the quality of the pastures or of the milk itself, than to the quantity of rennet with which it has there been customary to coagulate the milk.

The Way in Which the Rennet Is Made

This, no less than its state of preservation and the quantity employed, may also influence the flavor or other qualities of the cheese. For instance, in the manufacture of a celebrated French cheese, that of Epoisse, the rennet is prepared as follows:—Four fresh calf-skins, with the curd they contain, are well washed in water, chopped into small pieces, and digested in a mixture of 5 quarts of brandy with 15 of water, adding at the same time 2½ lbs. of salt, half an ounce of black pepper, and a quarter of an ounce each of cloves and fennel seeds. At the end of six weeks, the liquor is filtered and preserved in well-corked bottles, while the membrane is put into salt-water to form a new portion of rennet. For making rich cheeses, the rennet should always be filtered clear.

On Mont d'Or, the rennet is made with white wine and vinegar. An ounce of common salt is dissolved in a mixture of half a pint of vinegar with 2½ pints of white wine, and in this solution a prepared goat's stomach or a piece of dried pig's bladder is steeped for a length of time. A single spoonful of this rennet is said to be sufficient for 45 or 50 quarts of milk. No doubt the acid of the vinegar and of the wine aid the coagulating power derived from the membrane.

The Way in Which the Curd Is Treated

It is usual in our best cheese districts, carefully and slowly to separate the curd from the whey, not to hasten the separation, lest a larger portion of the fatty matter should be squeezed out of the curd, and the cheese should thus be rendered poorer than usual. But in some places, the practice prevails of washing the curd with hot water, after the whey has been partially separated from it. Thus at Gouda in Holland, after the greater part of the whey has been gradually removed, a quantity of hot water is added, and allowed to remain upon it for at least a quarter of an hour. The heat makes the cheese more solid and causes it to keep better.

In Italy, the pear-shaped *caccio-cavallo* cheeses and the round *palloni* cheeses of Gravina, in the Neapolitan territory, are made from curd, which after being scalded with boiling whey, is cut into slices, kneaded in boiling water, worked with the hand till it is perfectly tenacious and elastic, and then made into shapes. The water in which the curd is washed, after standing twenty-four hours, throws up much oily matter, which is skimmed off and made into butter.

The Separation of the Whey

Is a part of the process, upon which the quality of the cheese in a considerable degree depends. In Cheshire, more time and attention is devoted to the perfect extraction of the whey than in almost any other district. Indeed, when it is considered that the whey contains sugar and lactic acid, which may undergo decomposition, and a quantity of rennet which may bring on fermentation, by both of which processes the flavor of the cheeses must be considerably affected, it will appear of great importance that the whey should be as completely removed from the curd as it can possibly be. To aid in effecting this, a curd-mill, for chopping it fine after the whey is strained off, is in use in many of the large English dairies, and a very ingenious and effectual pneumatic cheese-press for sucking out the whey, was lately invented.

But the way in which the whey is separated is not a matter of indifference, and has much influence upon the quality of the cheese. Thus, in Norfolk, according to Marshall, when the curd is fairly set, the dairy-maid bares her arm, plunges it into the curd, and with the help of her wooden ladle, breaks up minutely, and intimately mixes the curd with the whey. This she does for ten or fifteen minutes, after which the curd is allowed to subside, and the whey is drawn off. By this agitation, the whey must carry off more of the butter and the cheese must be poorer.

In Cheshire and Ayrshire, the curd is cut with a knife, but is gently used and slowly pressed till it is dry enough to be chopped fine, and thus more of the oily matter is retained. On the same principle, in making the Stilton cheese, the curd is not cut or broken at all, but is pressed gently and with care till the whey gradually drains out. Thus the butter and the curd remain intermixed, and the rich cheese of Stilton is the result. Thus, while it is of importance that all the whey should be extracted from the curd, yet the quickest way may not be the best. More time and care must be bestowed in order to effect this object, the richer the cheese we wish to obtain.

The quality of the milk or of the pastures, may often be blamed for the deficiencies in the richness or other qualities of cheese, which are in reality due to slight but material differences in the mode of manufacturing it.

The kind of salt used, is considered by many to have some effect upon the taste of the cheese. Thus the cheese of Ge rome, in the Vosges, is supposed to derive a peculiar taste from the Lorena salt with which it is cured. In Holland, the efficacy of one kind of salt over another for the curing of cheese is generally acknowledged.

The Mode in Which the Salt Is Applied

In making the large Cheshire cheeses, the dried curd, for a single cheese of sixty pounds, is broken down fine and divided into three equal portions. One of these is mingled with double the quantity of salt added to the others, and this is so put into the cheese-vat as to form the central part of the cheese. By this precaution, the after-salting on the surface is sure to penetrate deep enough to cure effectually the less salted parts.

In the counties of Gloucester and Somerset the curd is pressed without salt, and the cheese, when formed, is made to absorb the whole of the salt afterwards through its surface. This is found to answer well with the small and thin cheeses made in those counties, but were it adopted for the large cheeses of Cheshire and Dunlop, or even for the pine-apple cheeses of Wiltshire, there can be no doubt that their quality would frequently be injured. It may not be impossible to cause salt to penetrate into the very heart of a large cheese, but it cannot be easy in this way to salt the whole cheese equally, while the care and attention required must be greatly increased.

Addition of Cream or Butter to the Curd

Another mode of improving the quality of cheese, is by the addition of cream or butter to the dried and crumbled curd. Much diligence, however, is required fully to incorporate these, so that the cheese may be uniform throughout. Still this practice gives a peculiar character to the cheeses of certain districts. In Italy, they make a cheese after the manner of the English, into which a considerable quantity of butter is worked; and the Reckem cheese of Belgium is made by adding half an ounce of butter and the yolk of an egg, to every pound of pressed curd.

Size of the Cheese

From the same milk, it is obvious that cheeses of different sizes, if treated in the same way, will, at the end of a given number of months, possess qualities in a considerable degree different. Hence, without supposing any inferiority, either in the milk or in the general mode of treatment, the size usually adopted for the cheeses of a particular district or dairy, may be the cause of a recognised inferiority in some quality, which it is desirable that they should possess in a high degree.

The Method of Curing

This has very much influence upon the quality of the cheese. The care with which they are salted, the warmth of the place in which they are kept during the first two or three weeks, the temperature and closeness of the cheese-room in which they are afterwards preserved, the frequency of turning, of cleaning from mould, and rubbing with butter; all these circumstances exercise a remarkable influence upon the after-qualities of the cheese. Indeed, in very many instances, the high reputation of a particular dairy district or dairy farm, is derived from some special attention to some or to all of these apparently minor points.

In Tuscany, the cheeses, after being hung up for some time at a proper distance from the fire, are put to ripen in an underground, cool, and damp cellar; and the celebrated French cheeses of Roquefort, are supposed to owe much of the peculiar estimation in which they are held, to the cool and uniform temperature of the subterranean caverns in which the inhabitants of the village have long been accustomed to preserve them.

Ammoniacal Cheese

The influence of the mode of curing, is shown very strikingly in the small ammoniacal cheeses of Brie, which are very much esteemed in Paris. They are soft unpressed cheeses, which are allowed to ripen in a room, the temperature of which is kept between 60° and 70° Fahrenheit, till they begin to undergo the putrefactive fermentation, and emit an ammoniacal odor. They are generally unctuous, and sometimes so small as not to weigh more than an ounce.

Inoculating Cheese

It is said that a cheese, possessed of no very striking taste of its own, may be inoculated with any flavor we approve, by putting into it with a scoop a small portion of the cheese which we are desirous that it should be made to resemble. Of course, this can apply only to cheeses otherwise of equal richness, for we could scarcely expect to give a Gloucester the flavor of a Stilton, by merely patting into it a small portion of a rich and esteemed Stilton cheese. [*Johnston and various other authorities.*]

Figure 19 is a self-acting cheese-press, light yet strong. The cheese itself gives a pressure of twelve times its own weight; and if this is insufficient, additional weight may be added as required.

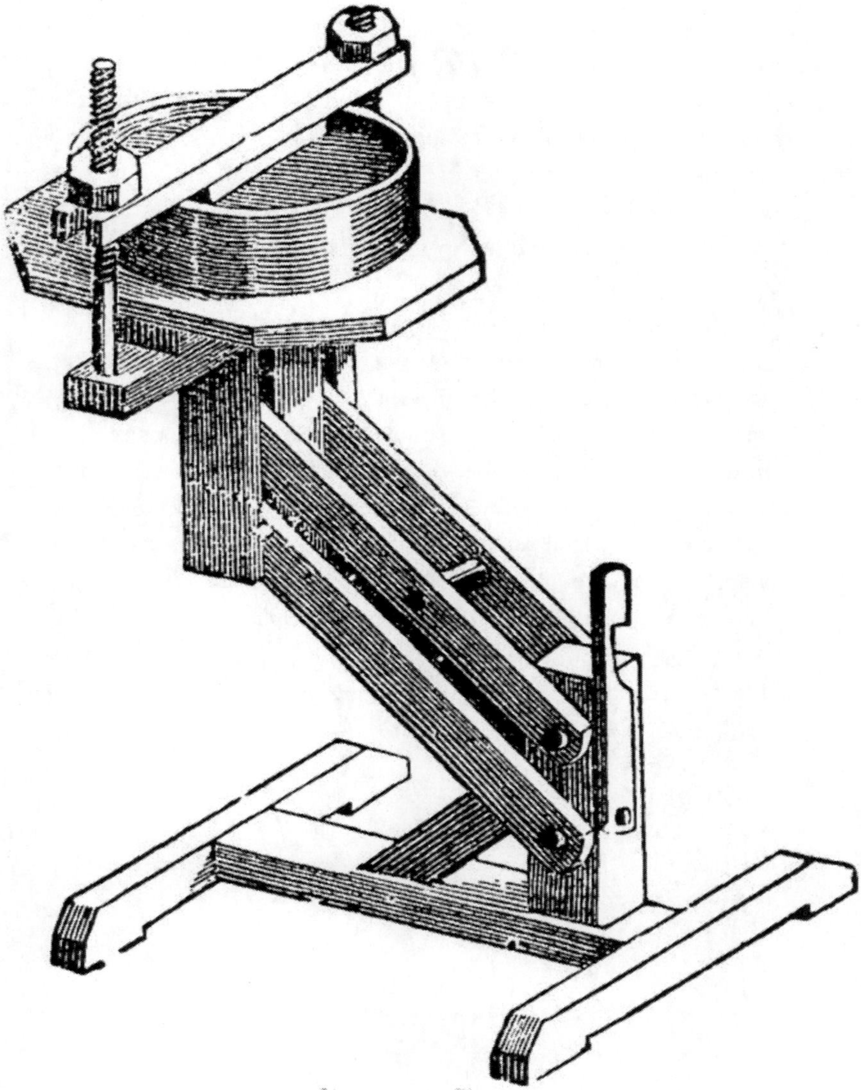

Figure 19. Cheese-Press.

The following statements were made by those receiving premiums from the New York State Agricultural Society:

> "Number of cows kept, eleven. Cheese made from two milkings, in the English manner; no addition made of cream. For a cheese of twenty pounds, a piece of rennet about two inches square is soaked about twelve

hours in one pint of water. As rennets differ much in quality, enough should be used to coagulate the milk *sufficiently* in about forty minutes. No salt is put *into* the cheese, nor any on the outside during the first six or eight hours it is pressed; but a thin coat of fine Liverpool salt is kept on the outside during the remainder of the time it remains in press. The cheeses are pressed forty-eight hours, under a weight of seven or eight cwt. Nothing more is required but to turn the cheeses once a day on the shelves."

"The milk is strained in large tubs over night; the cream stirred in milk, and in the morning strained in same tub; milk heated to natural heat; add color and rennet; curd broke fine and whey off, and broke fine in hoop with fast bottom, and put in strainer; pressed twelve hours; then taken from hoop, and salt rubbed on the surface; then put in hoop, without strainer, and pressed forty-eight hours; then put on tables, and salt rubbed on surface, and remain in salt six days, for cheese weighing thirty pounds. The hoops to have holes in the bottom; the crushings are saved, and set, and churned, to grease the cheese. The above method is for making one cheese per day. As in butter-making, the utmost cleanliness is required in every part of the cheese-making premises."

Chapter 4

SHEEP

With the exception of the dog, there is no one of the brute creation which exhibits the diversity of size, color, form, covering, and general appearance, which characterizes the sheep, and none which occupies a wider range of climate, or subsists on a greater variety of food. In every latitude between the equator and the arctic, he ranges over the sterile mountains, and through the fertile valleys. He feeds on almost every species of edible forage, the cultivated grasses, clovers, cereals and roots; he browses on aromatic and bitter herbs; he crops the leaves and bark from the stunted forest shrubs, and the pungent, resinous evergreens. In some parts of Norway and Sweden, when other resources fail, he subsists on fish or flesh during their long and rigorous winters, and if reduced to necessity, he eats his own wool.

He is diminutive like the Orkney, or massive like the Teeswater. He is policerate or many horned; he has two large or small spiral horns like the Merino, or is polled or hornless like the mutton sheep. He has a long tail like our own breeds; a broad tail, like many of the eastern; or a mere button of a tail, like the fat-rumps, discernible only by the touch. His coat is sometimes long and coarse, like the Lincolnshire; short and hairy, like those of Madagascar; soft and furry, like the Angola; or fine and spiral, like the silken Saxon. His color, either pure or fancifully mixed, varies from the white or black of our own country, to every shade of brown, dun, buff, blue, and gray, like the spotted flocks of the Cape of Good Hope and other parts of Africa and Asia. This wide diversity is the result of long domestication, under almost every conceivable variety of condition.

USES

Among the antediluvians, sheep were immolated for sacrificial offerings, and their fleeces probably furnished them with clothing. Since the deluge, their flesh has with all nations been used as a favorite food for man. By many of the rude, roving nations of the East, they are employed in carrying burdens.

Their milk is generally used by the uncivilized, and to some extent by the refined nations of Europe, not only as a beverage, but for making into cheese, butter, and curds. Job refers to its use, as do Isaiah and other of the Old Testament writers. Most of the Greek and Roman authors describe its general use and manufacture. The ewe's milk scarcely differs in appearance from that of the cow, but is generally thicker, and yields a pale, yellowish butter, that is always soft and soon becomes rancid. Culley remarks, "the cheese from their milk is exceedingly pungent, and for that reason is preferred by many to that from the cow." In Wales, the milk is mixed with that of the dairy, and makes a tart, palatable cheese. We have never seen it appropriated for dairy purposes in the United States, except by a few Welsh and Highland emigrants. The sheep is frequently employed in the dairy regions of this country, at the tread-mill or horizontal wheel, to pump the water, churn the milk, or perform other light domestic work.

The dignity and importance of the shepherd's vocation have ever been conspicuous. Abel, the supposed twin-brother of the first-born of the human race, was a "keeper of sheep;" and from this, it may be fairly inferred, that there is no animal which has so long been under the immediate control of man. Abraham and his descendants, as well as most of the ancient patriarchs, were shepherds. Job had 14,000 sheep. It is said of Rachel, the favored mother of the Jewish race, "she came with her father's sheep, for she kept them." The seven daughters of the priest of Midian, "came and drew water for their father's flocks." Moses, the statesman and lawgiver, who "was learned in all the wisdom of the Egyptians, kept the flocks of Jethro, his father-in-law;" and David, the future monarch of Israel, the hero, poet, and divine, was a keeper of sheep. It was to shepherds, while "abiding in the field, keeping watch over their flocks by night," that the birth of the Saviour was announced. The root of the Hebrew name for sheep, signifies fruitfulness, abundance, plenty; as indicating the blessings they were destined to confer on the human race. With the sacred writers, they were the chosen symbol of purity and the gentler virtues; they were the victims of propitiatory sacrifices; and finally, they became the type of redemption to fallen man. These may not be considered accidental allusions in a book, whose every feature is full of design.

Nor has the sheep been less the subject of eulogy and attention with profane writers. Among these, Homer and Hesiod, Virgil and Theocritus, introduced them with evident delight in their pastoral themes; while their heroes and demigods, Hercules and Ulysses, Æneas and Numa, carefully perpetuated them throughout their regal domains.

In modern times, they have commanded the attention of the most enlightened nations; and their prosperity has in no instance been independent of those useful animals, wherever wool and its manufactures have been regarded as essential staples. Spain and Portugal, for more than two centuries, were the most enterprising nations of Europe, and during that period, they excelled in the production and manufacture of wool. Flanders, for a time, was before England in the perfection of the arts and the enjoyments of life, and England then sent the little wool she raised to that country to be manufactured. Her politic sovereigns soon found this a losing game, and offered large bounties for the importation of artists and machinery. By a systematic and thorough course of legislation, which looked to the utmost protection and augmentation of wool and woollens, she has carried their production beyond anything the world has ever seen. The small islands of Great Britain and Ireland, in addition to the support of their 26,000,000 of people, 15,000,000 of cattle, 2,250,000 horses,

18,000,000 swine, and innumerable smaller domestic animals, maintain over 40,000,000 sheep, worth $250,000,000; and besides manufacturing nearly all their fleeces, annually import nearly an equal amount from abroad. The sumptuary law for burying the dead in woollen, still occupies its place in their statute book. And beyond all question, England is the leading power of the nineteenth century, in the combination of all those qualities which constitute national greatness, civilization, and strength.

VARIETIES

Naturalists have divided the wild sheep into four varieties. The *Musimon*, (*Ovis Musimon*,) inhabiting Corsica, Sardinia, and other islands of the Mediterranean, the mountainous parts of Spain and Greece, and some other regions bordering upon that inland sea, have been frequently domesticated and mixed with the long-cultivated breeds.

The *Argali*, (*O. Ammon*,) ranges over the steppes, or elevated plains of Central Asia, northward and eastward to the ocean. They are larger, more hardy, and more untameable than the Musimon.

The *Rocky Mountain Sheep*, (*O. Montana*,) frequently called the *Big-horn* by our western hunters, is found on the prairies west of the Mississippi, and throughout the wild mountainous regions, extending through California and Oregon to the Pacific. They are larger, but in other respects resemble the Argali, of which they are probably descendants, as they could easily cross upon the ice at Behring's Straits, from the northeastern coast of Asia. Like the Argali, when caught young they are easily tamed; but we are not aware that they have ever been bred with the domestic sheep. Before the country was overrun by the white man, they probably inhabited the region bordering on the Mississippi. Father Hennepin, a French Jesuit, who wrote nearly two hundred years ago, often speaks of meeting with goats in his travels through what is now the territory embraced by Illinois and Wisconsin. The wild, clambering propensities of these animals, occupying the giddy heights far beyond the reach of the traveller, and the outer coating of hair (supplied underneath, however, with a thick coating of soft wool) gives to them much of the appearance of that animal. In summer they are generally found single; but when they descend from their isolated, rocky heights in winter, they are gregarious, marching in flocks under the guidance of leaders.

The *Bearded Sheep of Africa* (*O. Tragelaphus*) inhabit the mountains of Barbary and Egypt. They are covered with a soft, reddish hair, and have a mane hanging below the neck, and large locks of hair at the ankle.

The Domesticated Sheep (*O. Aries*)

Embraces all the varieties of the subjugated species. Whether they have descended from any one of the wild races, is a question yet undetermined among naturalists; but however this may be, many of the varieties apparently differ less from their wild namesakes than from each other.

The *fat-rumped* and *the broad-tailed sheep* are much more extensively diffused than any other. They occupy nearly all the southeastern part of Europe, Western and Central Asia, and Northern Africa. They are supposed to be the varieties which were propagated by the patriarchs and their descendants, the Jewish race. This is inferred from various passages in the Pentateuch, Exodus xxix. 22; Leviticus iii. 9; viii. 25; ix. 19, and some others, where "the fat and the rump" are spoken of in connection with offerings, in which the fat was always an acceptable ingredient. Dr. Boothroyd renders one of the foregoing passages, "the large, fat tail entire, taken clear to the rump." It is certain this variety gives indisputable evidence of remote and continued subjugation. Their long, pendent, drowsy ears, and the highly artificial posterior developments, are characteristic of no wild or recently-domesticated race.

This breed consists of numerous sub-varieties, differing in all their characteristics of size, fleece, color, &c., with quite as many and marked shades of distinction as the modern European varieties. In Madagascar, they are covered with hair; in the south of Africa, with coarse wool; in the Levant, and along the Mediterranean, the wool is comparatively fine; and from that of the fat-rumped sheep of Thibet the exquisite Cashmere shawls are manufactured. Both rams and ewes are sometimes bred with horns, and sometimes without, and they exhibit a great diversity of color. Some yield a carcass of scarcely 30 lbs., while others have weighed 200 lbs. dressed. The tail or rump varies greatly, according to the purity and style of breeding; some are less than one-eighth, while others exceed one-third the entire dressed weight. The fat of the rump or tail is considered a great delicacy, and in hot climates resembles oil, and in colder, suet.

The broad-tailed sheep were brought into this country, about 50 years since, by Commodore Barron and Judge Peters, and bred with the native flocks. They were called the Tunisian mountain sheep. Some of them were subsequently distributed by Col. Pickering, of Massachusetts, among the farmers of Pennsylvania; and their mixed descendants were highly prized as prolific and good nurses, coming early to maturity, attaining large weights, of a superior quality of carcass, and yielding a heavy fleece of excellent wool. The principal objection brought against them, was the difficulty of propagation, which always required the assistance of the shepherd. The lambs were dropped white, red, tawny, bluish, or black; but all excepting the black, grew white as they approached maturity, retaining some spots of the original color on the cheeks and legs, and sometimes having the entire head tawny or black. The few which descended from those originally imported into this country, have become blended with American flocks, and are now scarcely distinguishable from them.

Native or Common Sheep of the United States

Strictly speaking, there are no sheep indigenous to North America, excepting the *Ovis Montana*, or Rocky Mountain sheep. Before the introduction of the improved European breeds, during the present century, our sheep were generally a hardy, long-legged, coarse, open-fleeced animal, which yielded, according to attention and feed, from 1½ to 4 lbs. of indifferent wool. We have seen numerous flocks within the last 20 years, of the pure-bred native, whose bellies were entirely destitute of wool, and sometimes the whole carcass was bare, excepting a mere strip or ridge like a mane, reaching from the head to the tail. The wool which was retained on the neck, back, and sides, was frequently matted almost as firmly as a leather apron; and that on the thighs, and sometimes on the sides, was often composed almost wholly of long hair.

Although indifferently formed in comparison with the best breeds of the present day, being thin in the breast and back, light quartered, and slow in coming to maturity, they yet possessed some good qualities. They were prolific, excellent nurses, tallowed well, and yielded good mutton. There were, occasionally, some smutty-nosed or brockle-faced sheep among them, distinguished by their additional size, superior merits, and courage. These were usually the leaders of the flock, in their marauding expeditions on their neighbor's domains; and in common with the others, they were eminently adapted to purvey for themselves on the frontier settlements. There were, besides, some black or dark chocolate-brown members in every flock, which were much valued by the thrifty housewife for their wool, which afforded an economical mixture for jackets, hose, and trousers, known as sheep's gray.

Our original stock were principally derived from England, where their counterparts may be seen at the present day, in the refuse breeds of that country. When these sheep were well selected and properly bred, there was rapid and satisfactory improvement, and from such flocks, mixed with some of the more recently improved varieties, have sprung many valuable animals.

There was but one exception to this general character of the native flocks, so far as our observation extended, which was a considerably numerous, and, probably, accidental variety, known as the *Otter breed*, or *Creepers*. These were an excessively duck-legged animal, with well-formed bodies, full chest, broad backs, yielding a close heavy fleece of medium quality of wool. They were deserved favorites where indifferent stone or wood fences existed, as their power of locomotion was absolutely limited to their enclosures, if protected by a fence not less than two feet high. The quality of their mutton was equal, while their aptitude to fatten was decidedly superior to their longer-legged contemporaries. They are probably now nearly or quite extinct.

An excellent variety was produced by General Washington, from a cross of a Persian ram, upon the Bakewell, which bore wool 14 inches in length, soft and silky, and admirably suited to combing. They were called the Arlington sheep, but they have long since become incorporated with the other flocks of the country.

The Merino

This is undoubtedly one of the most ancient race of sheep extant. The loose descriptions and indefinite generalities of the ancient writers, leave much to conjecture on this point; yet we have a few passages from Pliny, Columella, and some other Roman authors, which leave little doubt that the Merino was bred in their age, and had even been introduced into Italy from Greece. It is a matter of history, that the Greeks had choice breeds of sheep at an early day, which they might have derived from Egypt, Tyre, and Asia Minor, as they were intimately connected in commerce with those countries, where the woollen manufacture early reached great perfection. It is supposed that the celebrated Argonautic expedition, in quest of the golden fleece, undertaken by the Greeks nearly 1300 years before Christ, resulted in procuring a valuable race of sheep from Colchis, in the Euxine.

However this may be, it is certain that when Augustus extended his peaceful sceptre over half the known world, the Romans were in possession of some flocks, bearing fleeces of exceeding fineness and beauty. They had been reared in the province of Apulia, on the southeast coast of Italy, and were called Tarentine, from Tarentum, the capital of the province. Here, then, may have been one branch of the Merino family.

Another is undoubtedly described by Pliny, who says, "the *red fleece of Bætica* was of still superior quality, *and had no fellow*." All the Spanish coast on the Mediterranean, of which Bætica formed a considerable part, comprising the modern Spanish provinces of Jaen, Cordova, Seville, Andalusia, and Granada, was early colonized by the enterprising Greeks;

and this *red fleece that had no fellow*, was probably introduced by them at an early day, and by their descendants had been carried to a still higher degree of perfection than that of Apulia. Columella, the uncle of the writer on agriculture, a wealthy emigrant to Spain from Italy, A. D. 30, carried with him some of the Tarentine sheep, and thus added to the fine-woolled sheep of Spain. These two ancient streams, united perhaps with a third from the more ancient stock of the Euxine, (for Strabo asserts that some of the finest-woolled sheep were brought from that region in his time, and sold for the enormous sum of $750,) flowed on in an uninterrupted current over that broad country, and brought down to modern times the unrivalled race of the Merino. The limited region of Italy, overrun as it repeatedly was by hordes of barbarians during and after the times of the late emperors, soon lost her pampered flocks; while the extended regions of Spain, intersected in every direction by almost impassable mountains, could maintain their more hardy race, in defiance of revolution or change.

Whatever distrust may be attached to these scraps of history, which apparently establish the remote antiquity of the Merino, this much is absolutely certain, that they are a race whose qualities are inbred, to an extent surpassed by no others. They have been improved in the general weight and evenness of their fleece, as in the celebrated flock of Rambouillet; in the uniformity and excessive fineness of fibre, as in the Saxons; and in their form and feeding qualities, in various countries; but there has never yet been deterioration either in quantity or quality of fleece or carcass, wherever transported, if supplied with suitable food and attention. Most sheep annually shed their wool if unclipped; while the Merino retains its fleece, sometimes for five years, when allowed to remain unshorn. This we conceive affords conclusive evidence of long-continued breeding among themselves, by which the very constitution of the wool-producing organs beneath the skin have become permanently established; and this property is transmitted to a great extent even among the crosses, thus marking them as an ancient and peculiar race.

The conquest by the Moors of a part of those fine provinces in the south of Spain, so far from checking, served rather to encourage the production of fine wool. They were not only enterprising, but highly skilled in the useful arts, and carried on extensive manufactories of fine woollen goods, which they exported to different countries. After their expulsion in the 15th century, by Ferdinand and Isabella, the Spaniards preserved these manufactures in part, and sedulously cherished their fine flocks; and knowing the incomparible advantage they had in them, their sovereigns, except in a few isolated instances, strictly prohibited their exportation.

Exportation of Merinoes from Spain

History asserts that Henry VIII of England, by permission of Charles V., imported 3,000 Spanish sheep; but of what kind is not mentioned, they having numerous varieties in Spain. If of the true Merino, it will explain the superior quality of the English middle-wools, the Ryeland, South Downs, and some others.

The first well-authenticated exportation of the Spanish Merino, was made to Sweden in 1723, by Alstroemer, which solved the problem of their capacity for sustaining their character, on rough fare and in a high northern latitude. Lasteyrie, who wrote fifty years after the experiment had been tried, speaks of their improvement, both in carcass and the quality and quantity of fleece.

The next exportation was made to Saxony, in 1765, and consisted of 105 rams and 114 ewes, but from what flocks they were taken, history nowhere mentions. A second exportation to that country was made in 1778, of 110 that were variously selected, from the best flocks in Spain. From these have descended the high-bred, silken-fleeced Saxons, whose wool stands confessedly without a rival.

In 1775, the Empress Maria Theresa imported 300 Merinoes into Germany, and placed them on the imperial farm in Hungary. In 1786, an importation was made into Denmark and her provinces; and again, in 1797, another flock of 300 was brought into the kingdom, and placed at Esserum, about eight leagues from Copenhagen. In 1786, 100 rams and 200 ewes were imported into Prussia, most of which were allowed to perish from neglect and disease; but their places were fully made up by later importations. The same year, 400 ewes and rams were selected from the choicest Spanish flocks, and placed on the royal farm of Rambouillet, in France, which laid the foundation of the celebrated flock which bears that name.

A small flock of inferior animals was clandestinely procured by George III., of England, in 1788, which attracted little attention. In 1791, a small but choice flock was presented to that monarch by the Cortes of Spain, which soon acquired high favor among many intelligent breeders. A part of these were kept pure, and their descendants furnished the superb flock of 700 Nigrettis, which procured for their owner, Mr. Trimmer, in 1829, the gold medal from the London Society of Arts. Others were mixed with different flocks in the kingdom, to the evident improvement of their fleeces.

The First Importation of Merinoes into the United States

Which resulted in the propagation of a pure breed,[2] was made by Chancellor Livingston, then minister at the court of Versailles, who sent two choice rams and ewes from the Rambouillet flock, in 1802, to Claremont, his country seat on the Hudson. In the latter part of the same year, Col. Humphreys, our minister in Spain, sent out nearly one hundred Merinoes, which were followed by more numerous flocks from the same and other sources. The largest importations of the Merino, however, were made through Mr. Jarvis of Vermont, in 1809, then U. S. Consul in Spain, and immediately thereafter. He first shipped, as he states, "200 Escurial, afterwards 1400 Paulars, 1700 Aqueirres, 100 Nigrettis, and about 200 Montarcos. 2700 Montarcos were sent out by a Spaniard and a Portuguese, and about 300 Guadaloupes by others; also 200 to 300 Paulars, by Gen. Downie, to Boston. Of the Montarco flock shipped by others, about 2500

[2] One or more pure Merinoes were imported into Massachusetts, in the latter part of the last century, by a citizen of that state, but they were soon mixed with other flocks, and resulted in the perpetuation of no distinct breed.

came to Boston, Providence, New York, and other ports. All were imported in the latter part of 1809 and '10, and early in 1811, and were the only Leonese Transhumantes, if we include Humphreys' and Livingston's, (which I have no doubt were of the same stock,) that were ever shipped to the United States."

Figure 20 is a spirited cut of a variety of the Merino without dewlap, and with a long and somewhat open fleece.

Figure 20. Merino Buck.

Varieties of the Spanish Sheep

Besides several other breeds of sheep in Spain, consisting of long, coarse wool, and that of a medium staple, embraced under the different names of *Chorinoes*, *Choaroes* or *Chunahs*, the Merino is distinguished by two general divisions; the *Transhumantes* or travelling, and the *Estantes* or stationary flocks. The former are subdivided, according to the Provinces they occupy, into Leonese, Segovian, and Sorian. Many of the Estantes were of the best quality in respect to carcass, constitution, and fleece; and such as were highly bred and in the hands of intelligent breeders, were not surpassed by any of the Spanish flocks. There were also many choice sheep among the Segovian and Sorian Transhumantes, but in general they were decidedly inferior to those of Leon. These last were universally regarded as the prime flocks of Spain. They comprised the Escurial, the Paular, the Nigretti, the Aqueirres or Muros, the Montarco, the Guadaloupe, Infantado, and some others.

There is much contradictory testimony as to the comparative merits of the last-mentioned flocks, as they were found in Spain; which is owing in part, doubtless, to the difference in the specimens subjected to examination. We subjoin some of the most reliable authorities on this subject.

M. Lasteyrie, who investigated this matter closely, says, "The Guadaloupe have the most perfect form, and are likewise celebrated for the quantity and quality of their wool. The Paular bear much wool of a fine quality, but they have a more evident enlargement behind the ears, and a greater degree of throatiness, and the lambs have a coarse hairy appearance, which is succeeded by excellent wool. The lambs of the Infantado have the same hairy coat when young. The Nigretti are the largest and strongest of all the travelling sheep in Spain."

Mr. Livingston says, "The Escurial is the most perfect of all the travelling flocks in Spain; the Guadaloupe for form, fineness and abundance of the fleece; the Paular with similar fleeces are larger bodied. Those of Castile and Leon have the largest, with the finest coat. Those of Soria are small, with very fine wool; and those also of Valencia, which do not travel, and like the last have fine wool, but of a very short staple."

Mr. Jarvis, who spent many years in Spain, under every advantage for studying them closely, and who also imported, and has since bred large numbers of them on his estate in Vermont, gives their characteristics with more particularity, and at much greater length: "The Paulars were undoubtedly one of the handsomest flocks in Spain. They were of middling height, round-bodied, well spread, straight on the back, the neck of the bucks rising in a moderate curve from the withers to the setting on of the head, their head handsome, with aquiline curve of the nose, with short, fine, glossy hair on the face, and generally hair on the legs, the skin pretty smooth, that is, not rolling up or doubling about the neck and body, as in some other flocks; the crimp in the wool was not so short as in many other flocks, the wool was somewhat longer, but it was close and compact, and was soft and silky to the touch, and the surface was not so much covered with gum. This flock was originally owned by the Carthusian friars of Paular, who were the best agriculturists in Spain, and was sold by that order to the Prince of Peace when he came into power.

The Nigretti flock were the tallest Merinoes in Spain, but were not handsomely formed, being rather flat-sided, roach-back, and the neck inclining to sink down from the withers; the wool was somewhat shorter than the Paular, and more crimped; the skin was more loose and inclined to double, and many of them were woolled on their faces and legs down to their hoofs. All the loose-skinned sheep had large dewlaps.

The Aqueirres were short-legged, round, broad-bodied, with loose skins, and were more woolled about their faces and legs than any other flock I ever saw; the wool was more crimped than the Paular, and less than the Nigretti, but was thick and soft. This flock formerly belonged to the Moors of Spain, and at their expulsion was bought by the family of Aqueirres. The wool in England was known as the Muros flock, and was highly esteemed. All the bucks of these three flocks had large horns.

The Escurials were about as tall as the Paulars, but not quite so round and broad, being in general rather more slight in their make; their wool was crimped, but not quite so thick as the Paular or Nigretti, nor were their skins so loose as the Nigretti and Aqueirres, nor had they so much wool on the face and legs.

The Montarco bore a considerable resemblance to the Escurials. The Escurial flock had formerly belonged to the crown, but when Philip II built the Escurial palace, he gave them to the friars, whom he placed in a convent that was attached to the palace, as a source of revenue. These four flocks were moderately gummed.

The Guadaloupe flock was rather larger in the bone than the two preceding, about the same height; but not quite so handsomely formed; their wool was thick and crimped, their skins loose and doubling, their faces and legs not materially different from the two latter flocks, but in general they were more gummed than either of the other flocks. In point of fineness, there was very little difference between these six flocks; and as I have been told by well-informed persons, there is very little difference in this respect among the Leonese Transhumantes in general. The Escurials, the Montarcos, and the Guadaloupes, were not, in general, so heavy-horned as the other three flocks, and about one in six of the bucks were without horns."

The Saxon

We have before seen, is one of the varieties of the pure-bred Merino, the foundation of which was laid by an importation of some of the choicest animals into Saxony, in 1765. The great care and attention bestowed upon these sheep by the Elector, the nobility, and the most intelligent farmers, soon carried them to a point of uniformity and excellence of fleece, never exceeded by the best of the original flocks. The breeders were selected with almost exclusive reference to the quality of the fleece. Great care was taken to prevent exposure throughout the year, and they were housed on every slight emergency. The consequence of this course of breeding and treatment has been to reduce the size and weight of fleece, and partially to impair that hardiness and vigor of constitution, which universally characterized the original Transhumantes. In numerous instances, this management resulted in permanent injury to the character of their flocks, which America has severely felt in several importations of worthless animals, and which a too great eagerness for improvement induced her flockmasters to use with the Spanish Merinoes and their descendants, as a means for this object, but which has resulted in the introduction of fatal diseases and serious deterioration in their flocks.

Figure 21. Saxon Ram.

The First Importation of Saxons into This Country

Was made in 1823, of four good rams, two of which went to Boston and the others to Philadelphia. The next was made the following year, and consisted of 75 rams and ewes, which were brought to Boston, and sold at public auction, and were afterwards scattered over the country. Another lot of 180 followed the next year, to the same place, and was sold in the same manner, but at an increased price, some selling as high as $450 each.

These prices excited the spirit of speculation, and the following year witnessed the importation of near 3,000, many of which were decidedly inferior. These were all thrown upon the market for the most they would command; and in many instances, the sales not half covering the cost of importation, the enterprise was abandoned as a speculation, or commercial operation.

The late Henry D. Grove, of Hoosic, New York, a native of Germany, and a highly intelligent and thoroughly bred shepherd, accompanied some of the best early importations to this country. He selected 105 choice animals for his own breeding, which he imported in 1827, and 70 more equally good, in 1828, and with these he formed the flock from which he bred to the time of his decease, in 1844.

Figure 22. Rambouillet Buck.

The average weight of fleece from the entire flock of Mr. Grove, nearly all of which were ewes and lambs, as stated by him to the writer, in 1842, was 2 lbs. 14 ounces, thoroughly washed on the sheep's back. This was realized after a short summer and winter's keep, when the quantity of hay or its equivalent fed to the sheep did not exceed by actual weight 1½ lbs. per day, except to the ewes, which received an additional quantity just before and after lambing. This treatment was attended with no disease or loss by death, and with an increase of lambs, equalling one for every ewe.

In a flock of pure Saxony sheep owned by Mr. Smith of Connecticut, as stated in a letter from the owner, published in the American Shepherd, 104 ewes raised 101 lambs, and yielded 341 lbs. of wool, which sold at 70 cents per lb. For the 18 months preceding, he lost but three animals out of 300, from ordinary casualties. But some flocks of pure Saxony do not, in good condition, average 2 lbs. per head.

A recent importation, (May, 1846,) made by Mr. Taintor of Connecticut, consisting of four bucks and four ewes, from the celebrated Saxon flock of Baron de Spreck, show a size and apparent vigor of constitution, equalling any of their Merino progenitors.

The Rambouillet Flock

This flock was founded in 1786, by Louis XVI., from a selection of 400 of the best Spanish sheep, which were placed on the royal farm at Rambouillet. These, like the Saxon, received all the attention which intelligence and wealth could bestow, and the consequence was soon manifest in their larger size, and the increased weight and uniformity in the fineness of their fleece; the last improvement being particularly evident, from the absence of the coarse wool, which in many cases infested the quarters; and the jarr or hair, which frequently abounds on the flanks, legs, and thighs of the original Merino.

Besides the crown flocks at Rambouillet, they are found in equal perfection on several other of the royal farms, especially those of Malmaison, Perpignan, Arles, Clermont, and some others. These flocks have been bred for hardy constitution, large carcass, and heavy fleece, and of as much fineness as is consistent with large weights, and as uniform in quality throughout, as possible. Mr. Gilbert, who was particularly familiar with them, says, "almost all the fleeces of the rams, from two years old and upwards, weigh (unwashed) from 12 to 13 lbs.; but the mean weight, taking the rams and the ewes together, has not quite attained to 8 lbs., after deducting the tags and the wool of the belly." The French pound is about one-twelfth heavier than the English; but from the general custom of folding the sheep in France, feeding them in fallows, and wintering them in houses, the fleece becomes very dirty. The loss in washing (fit for manufacturing) is about 60 per cent., so that the clean fleece of the ram will average about 6 lbs., and that of the whole flock something under 4 lbs.

The First Importation of the Rambouillets to This Country

This was in 1801, by M. Dellesert, of Paris, for M. Dupont, then in New York, and consisted of four choice rams, only one of which, Don Pedro, reached this country. He was used among the native ewes near Kingston, N. Y., for three years, and then transferred to Delaware, where he effected great improvement among the native flocks. The second was that made by the late Chancellor Livingston, before alluded to. There was another in 1840, by Mr. Collins, of Connecticut, comprising 30 select ewes and 2 rams. All these sheep possessed the characteristics peculiar to the variety as described.

A still more recent importation has been made by Mr. Taintor, at the same time with that of the Saxons previously mentioned, consisting of 23 ewes and 3 bucks, variously selected from the choicest flocks of the descendants of the Spanish Merino. The rams, though young, are the most promising animals of their breed, and when full grown, will weigh from 225 to 250 lbs. each. The sire of one was sold the past season for $500. He sheared 23 lbs. of unwashed wool. The ewes measured after they were shorn, from 25½ to 29 inches in height over the withers. The height of the under side of their bodies from the ground, was from 9½ to 12 inches, which is in the proportion of good American Merino sheep. Their weights varied, after shearing, from 124 to 153 lbs.

Some of them were quite thin in flesh, the largest especially, which, if in fine condition with her fleece on, would weigh at least 200 lbs. The following is the weight of their fleeces unwashed. The scales used did not mark less than one-quarter of a pound, which will account for the absence of odd ounces.

No.	17	13	lbs.
"	27	15	"
"	64	16¾	"
"	71	14½	"
"	84	16½	"
"	87	16¼	"
"	94	17	"
"	100	12¼	"
"	109	17	"
"	110	17	"
"	117	16¾	"
"	118	15¾	"
"	133	14¾	"
"	195	13½	"

It was the unbiased opinion of several wool-dealers present, that the shearing above would yield at least 35 lbs. of *cleansed* wool, fitted for manufacturing without further loss, out of every 100 lbs. shorn. The fourteen ewes yielded 216 lbs. unwashed, which would be equivalent to 75 lbs. 10 oz. thoroughly cleansed, or an average of 5 lbs. 6 oz. per head. One-third may be safely added to bring this up to *clean washed*. This would make the average, as wool-growers usually dispose of their fleeces, 7 lbs. 3 oz. per head, a yield totally unprecedented in this country.

The Progress of the Merino in the United States

Though reaching back but half a century, the Merino flocks of this country have been very fluctuating as to their value, increase, and improvement. When first introduced, they were viewed with distrust by the majority of our farmers; and it was not till after several years' experience of their paramount merits, that they were generally disseminated. But the confidence of our flock-masters having once been secured, it has never been withdrawn, and they have ever since been cherished favorites. The prices for choice Merinoes rapidly increased after their character was fully established, and Livingston states the average price for rams, in 1811, at $1,000, and some were sold at a much higher rate.

This was the period of the embargo, when our infant manufactures were just starting into life; and being followed by war with the greatest commercial nation of the world, we were thrown entirely on our own resources for the supply of our woollen and other fabrics, and wool and sheep maintained their full value till after the return of peace, in 1816. The flooding of our country with foreign goods, under low duties, which succeeded this event, either broke down or effectually paralyzed our woollen manufactures; and wool, of course, felt the full weight of this crushing influence. The Merino rapidly declined in value, till its price nearly approximated to that of the native sheep. Their merits had, however, become so conspicuous, that the low prices produced a more general diffusion, and they and their crosses were thus sown broadcast over the country.

The introduction of the Saxons, in great numbers, in 1826, many of which were excessively diminutive and diseased, and their indiscriminate use with our pure-bred Merinoes, was a serious interruption to the career of improvement in many of our flocks. Their mixture with the best Saxons was no further detrimental, than to reduce the quantity of fleece, and, to a certain extent, lessen the peculiar hardiness of the original Transhumantes, which had been fully preserved by their descendants in this country. The use of well-selected Saxon rams with Merino flocks, was extensively practised, and it is still persisted in by many intelligent flock-masters, after twenty years' experience, who are satisfied that they find it for their interest to continue this style of breeding. The animals being smaller, consume less; and they probably produce a quantity of wool in proportion to their food, which, from its improved and uniform quality, commands a higher price in the market. Wherever they are not sufficiently hardy, they can be bred back towards the Spanish Merino standard, by the use of some of the stouter rams. Their natures are intrinsically the same. They are only divergent streams from the same original fountain, and when again united, they readily coalesce and flow onwards, without violence or disorder.

The Merino, as might reasonably have been anticipated, when properly managed, has improved from a variety of causes. Though kept scrupulously pure in Spain, they were seldom bred with that refinement of taste or nice judgment, which distinguishes the accomplished modern breeder. Their management was too entirely intrusted to ignorant shepherds or careless agents, to secure that close attention which is essential to improvement. The sheep had to perform a journey of several hundred miles twice in a year, to and from their distant Sierras; and it was absolutely essential that strong animals should be selected for breeding; and to secure this object, those were frequently used which were deficient in the most profitable qualities. They were also closely bred in-and-in, seldom or never departing from a particular flock to procure a fresh cross. Their wild, nomadic life, approaching nearly to that of their natural state, and their peculiarly healthful pasturage, alone prevented a serious deterioration from this cause.

When brought into the United States, the flocks were soon mingled with each other, and for many years past, probably, not an unmixed descendant of any distinct original flock could be traced. Abundance of appropriate food has been given them, without the labor of long and fatiguing journeys; and lastly, there has been much care used in the selection of the most profitable animals for breed. The spirit of improvement has been recently awakened to this important branch of American husbandry, and as we already have all the elements within ourselves for its attainment, if not arrested by any untoward national policy, it will soon result in giving us numerous flocks of as choice sheep as the world affords.

Peculiarities of the Merino

The prominent peculiarities of the Merino, are the abundance and fineness of its fleece; the tenacity with which it is held; its crimped or spiral form; its felting properties; and the excessive quantity of yolk, giving to it that softness which distinguishes it from all others. Their large horns are common to several other varieties. Their hoofs are sometimes singularly long, reaching 8 or 10 inches when allowed to grow. The horns, hoofs, and wool scarcely differ in their chemical constituents; and the peculiar development of the two former, is justly considered as an additional evidence of their wool-bearing properties.

The yolk, in most of the sheep, forms, with the dust which adheres to it, a firm crust on the exterior, and together with the compactness of the fleece, it offers considerable resistance to the open hand on being pressed, giving the impression of rigidity. This outer covering repels the rain, the snow, and the wind like a coat of mail, thus fitting the Merino to endure exposure beyond any other sheep. On opening the crust, the wool is found of a brilliant, golden hue, sparkling with yolk, and firmly held together in masses, hardly distinguishable from the cocoon of the silk-worm. The wool closely covers every part of the body, and frequently the entire legs and head, excepting a part of the face. Still another peculiarity of the Merino is its longevity. They attain a great age when properly managed, and, in healthy localities, sometimes breed till 20 years of age.

The Merino may be described, generally, as a small-boned, closely made, medium-sized sheep, varying from 80 lbs. of live weight, for a small ewe, to 160 lbs. for good-sized wethers and rams, in ordinary condition. They are light in the shoulders and chest, and altogether are more deficient in form than the best mutton sheep. This apparent difference is materially lessened when both are denuded of their fleece; as the longer pile of the latter covers defects, which would manifest themselves under the closer covering of the Merino. Yet, with this seeming deficiency, Young found, in feeding the Merino and Bakewell, that the latter ate the most, and gained the least, in the ratio of two to three. We give the statement as we find it, though it apparently contravenes a fundamental principle, which a knowledge of all the circumstances of the trial, the peculiarities of the particular animals, &c., might probably explain satisfactorily.

The mutton is excellent, and it is probably not surpassed by that of any other sheep. Lord Somerville claims it as a rule, that the quality of the flesh in each class of sheep follows that of the wool, and that the flesh of the short and fine-woolled sheep is closer in the grain and more highly flavored than the long-woolled. Sir Joseph Banks says, the London butchers, after having some of the Merinoes, eagerly sought for more, from its popularity with their best customers; and it is certain that the flavor of our mountain-fed Merino does not suffer in comparison with the choicest breeds.

Breeding Merinoes

The general principles of breeding cattle and sheep, as laid down by the most approved authorities, must be taken with some exceptions, when applied to the Merino. Good form and feeding qualities are desirable in this breed, but they are not as essential as with the others. *Wool* is the great object, and if this be sufficiently fine, even, and abundant, something may be abated in the perfection of form. Early maturity, so much sought after in the mutton sheep, cannot be reconciled with the great longevity, and the prolonged productive powers of the Merino. We must content ourselves, therefore, with slowly engrafting such improvements on the breed, as can be effected without prejudice to his other good qualities, and look to his crosses with others for such qualities as are irreconcilable with his nature.

It is considered indispensable to the improvement of the Merino, that it be not bred too young. A vigorous ewe may bring her first lamb at two years old, but it is better that it be deferred till three. The ram should never be used till his second year, and then but sparingly. From 2½ to 6 years old is deemed the most vigorous age, though many may be safely used till 8 or 10, and occasionally later. Both ewes and rams have been known to breed till 20 years old.

The ram should be large, stout, and well made, carrying his weight as compactly as possible. The nose should be convex; the face covered with a soft velvety hair; the eye lively and prominent; the veins near the lachrymal glands, of a clear red; the horns rough; short neck; pendent dewlap not objectionable; full chest; broad shoulders; broad, level back; large quarters; tail large and well set up; good legs, and sound hoofs; with a firm, easy, regular gait; the head carried high, with a look of boldness and decision, without in any degree approaching to wildness or ferocity.

The ewe should possess these characteristics generally, with such modifications as are suited to the sex. Great care should be taken to breed from such as are most perfect in all the essential points of constitution, form, and size; and weight, uniformity, and fineness of fleece. The closest observation is requisite, to select the best in all respects.

In-and-in breeding should be avoided where practicable, which can be done where there is a careful registry of the sheep through successive generations. Excessive use of rams can never be permitted without decided injury to them and their progeny. In Spain, four rams are supplied to every hundred ewes. This limited number is proper enough, where they undergo so much fatigue in travelling, and kept too, as they are, entirely on grass. But if moderately grained before and during their use, and especially if kept up, and allowed to serve the ewes once only, as they come in heat, this number may be largely increased. A vigorous ram will suffice for 35 to 40 ewes, when running with the flock; yet his powers would not be more taxed by double or even treble this number, if admitted to each but once. Bread is a convenient food for the ram while running with the sheep. If he is gentle, which he should always be, he will come up readily and eat from the hand, without exciting the attention of the other sheep, which crowd, and not unfrequently injure each other when grain is placed before him; or he may be stabled at night and fed with grain.

If young ewes have stolen lambs, they should be taken away immediately after yeaning, and the nourishment supplied to the lamb from the milk of a cow. The tax of nursing is nearly equal to that of gestation, and farther injury to the dam may be avoided by this practice. Merino ewes have had the reputation of being indifferent nurses in Spain. This is owing to their fatigue in travelling, and scanty pasturage, rather than to any constitutional deficiency. It is a frequent practice there, to kill a part of the lambs, and put one on to two ewes. This has never been found necessary in the countries where they have been transplanted, as generous feed for the dams has invariably been found entirely adequate to their support of the young.

The Localities in Which Merino Sheep Can Be Profitably Kept in the United States

Are wherever the pastures are sweet and dry; the climate not excessively hot; and the land not too valuable for other purposes. Wool is generally the great object in the sheep husbandry of this country, and when sheep farms are remote from the large markets, the Merino will make much the most profitable returns. In the neighborhood of cities, where large and fat sheep and early lambs bear a high price, the mutton sheep may be substituted.

The South Down

This valuable sheep has been known and bred for a long time on the chalky downs of England, where it has always maintained the character of a hardy animal, yielding a medium quality of wool, and furnishing mutton of a superior flavor. It was not, however, till within the last 70 years, that any considerable attention was devoted to its improvement. Since that period, its fine points have been remarkably developed, which is shown in its improved size and form, and its early maturity and productiveness. The late Mr. John Ellman, of England, was the first who took them thoroughly in hand; and so eminent was his success, that he founded a flock which has been the source whence all the best blood has been since derived.

Figure 23. South Down Buck.

The Form and Characteristics of the South Downs

His criteria of a good South Down, are as follows:—"The head small and hornless; the face speckled or gray, and neither too long nor too short. The lips thin, and the space between the nose and the eyes narrow. The under jaw, or chap, fine and thin; the ears tolerably wide, and well covered with wool, and the forehead also, and the whole space between the ears well protected by it, as a defence against the fly. The eye full and bright, but not prominent. The orbits of the eye—the eye-cap, or bone,—not too projecting, that it may not form a fatal obstacle in lambing. The neck of a medium length, thin towards the head, but enlarging towards the shoulders, where it should be broad and high, and straight in its whole course above and below. The breast should be wide, deep, and projecting forwards between the fore-legs, indicating a good constitution, and a disposition to thrive. Corresponding with this, the shoulders should be on a level with the back, and not too wide above; they should bow outward from the top to the breast, indicating a springing rib beneath, and leaving room for it. The ribs coming out horizontally from the spine, and extending far backward, and the last rib projecting more than the others; the back flat from the shoulders to the setting on of the tail; the loin broad and flat; the rump long and broad, and the tail set on high and nearly on a level with the spine. The hips wide; the space between them and the last rib on either side as narrow as possible, and the ribs, generally, presenting a circular form like a barrel. The belly as straight as the back. The legs neither too long nor too short. The fore-legs straight from the breast to the foot; not bending inward at the knee, and standing far apart both before and behind; the hocks having a direction rather outward, and the twist, or the meeting of the thighs behind, being particularly full; the bones fine, yet having no appearance of weakness, and of a speckled or dark color. The belly well defended with wool, and the wool coming down before and behind to the knee, and to the hock; the wool short, close, curled, and fine, and free from spiry projecting fibres."

Figure 24. South Down Ewe.

Other breeders have commenced where Ellman left off, and have apparently pushed their improvement to its utmost capacity; and especially has this been done by Messrs. Grantham and Webb, the latter of whom, while preserving all the essential merits of the sheep, has carried the live weight of breeding rams to 250 lbs., and well-fattened wethers to 200 lbs. dressed weight. Many of the choicest animals have been imported into this country, and they are now to be found, in limited numbers, in almost every state of the Union.

The wool was formerly short, and used only for cloths, flannels, &c. It has been considerably lengthened in many of the late flocks, and with the improvements in the combing machinery, is now much used in England as a combing wool. The quantity produced is nearly equal to that of the Merino flocks when well kept, varying, according to the size and style of breeding, from 3 to 4 lbs. of clean washed wool, which in quality does not differ materially from half-blood Merino, and sometimes rather exceeds it. The larger animals, of course, produce fleeces of much greater weight, sometimes reaching to 8 or 9 lbs. The South Down will subsist on short pasture, but well repays full feeding. It attains early maturity, is hardy and prolific, frequently producing two at a birth. Like all highly-improved English breeds, it is not a long-lived sheep. It may be considered in its prime at three. The wethers may be fattened at 18 to 30 months, and the ewes at 3 to 5 years, when first required as breeders. The last are sometimes allowed to come in with a lamb at a year, but they cannot be sustained in vigor if put to breeding before two.

The Cheviot

Is thus described by Blacklock: "They have a bare head, with a long jaw and white face, but no horns. Sometimes they have a shade of gray upon the nose, approaching to dark at the tip; at others, a tinge of lemon color on the face, but these markings scarcely affect their value.

The legs are clean, long, and small-boned, and covered with wool to the hough, but there is a sad want of depth at the breast, and of breadth both there and on the chine. A fat carcass weighs from 12 lbs. to 18 lbs. per quarter, and a medium fleece about 3 lbs. The purest specimens of this breed are to be found on the Scotch side of the Cheviot hills, and on the high and stony mountain farms which lie between that range and the sources of the Teviot. These sheep are a capital mountain stock, provided the pasture resembles the Cheviot hills, in containing a good proportion of rich herbage."

They are eminently adapted to high lands and a severe climate, though less so than the Black-faced or Heath sheep of Scotland. They have become an American sheep, by their repeated introduction into this country. A late importation of several choice sheep was made by Mr. Carmichael of New York. The wool on these is from 5 to 7 inches long, coarse, but well suited to combing. Like the Downs, it has heretofore been classed among the middle-wools, but these specimens would seem to indicate that they are verging towards the long-wools.

The Bakewell or Leicester, the Cotswold and Lincolnshire

Possess several qualities in common, and it is only a practiced eye that can readily detect the difference. This resemblance arises from a recent, common origin. They are all large and hornless; of a pure white; with long, coarse, and heavy fleeces; excellent mutton sheep; coming early to maturity, and capable of carrying enormous quantities of fat. There have been from time immemorial numerous flocks of these large, coarse-woolled sheep, existing in certain parts of England, under a variety of names, and partaking of some slight peculiarity of features, according to the district in which they are bred. Thus, besides those above-mentioned, there were the Teeswater, the Romney-Marsh, the Kentish, the Bampton, the Exmoor, &c., all of which were deficient in form, slow-feeders, and late in coming to maturity.

Improvement of the Long-Wools

The late Robert Bakewell first commenced a decided improvement with the Leicesters, nearly a century since. He began by selecting the choicest sheep in England, which possessed the essential qualities; and by judicious feeding and management throughout, he soon brought them up to a character widely differing from the original with which he started. So eminent was his success, that in 1787 he let three rams, for a single season, for 1250 pounds, (about $6,200,) and was offered 1050 pounds (about $5,200) for 20 ewes. Soon after this, he received the enormous price of 800 guineas, or $4,000, for the use of two-thirds of a ram for one season, reserving the other third for himself.

Figure 25. Long-wool Buck.

He reduced the bone and offal or worthless parts of the carcass, and increased the weight of the valuable parts, and especially their tendency to fatten and early maturity. This was effected mainly by a nice discrimination, which has probably never been surpassed, if it has ever been equalled. He selected medium sizes for the breed, with as much evenness and perfection of form as possible, for he found that excellence and profitable feeding qualities were seldom connected with extra size, large bones, or imperfect form. He also observed the disposition to fatten in individuals, and used only such as were conspicuous in this respect. He relied more than all upon their quality of *handling* well, depending even more upon the elastic, mellow touch, than upon the most symmetrical figure. He used only the choicest rams, a little under size, while the ewes were of full medium weight. The progeny were pushed with a full supply of nutritious food, and systematically brought to early maturity.

Connected with this, was his practice of *in-and-in breeding*, or breeding the parent upon the progeny, for several successive generations, which had the tendency still further to refine the bone and offal, and impress most effectually the desirable characteristics of the race. It is even credibly asserted, that he produced rot in such of his fattening sheep as he wished to mature early for the shambles, as in the first stages of that loathsome disease the fat-secreting organs accomplish their office more rapidly than in a state of perfect health; and it at least secured them against breeding when they left his own hands. It is certain, that Bakewell carried his refining system to such an extent, as partially to destroy the procreative powers; and he was subsequently obliged to introduce new animals, to reinvigorate and continue his flock.

The general system of Bakewell, however, was attended with complete success. He produced a race of animals, not only far beyond what England had ever before seen, but which, in all the qualities he endeavored to establish, have not been since exceeded; and his *improved Leicesters* have come down to the present day as perfect as he left them, showing conclusively, that he not only formed, but stamped the peculiarities of the breed, with a permanence which yet bears witness to his genius. One of these attained the enormous live weight of 368 lbs., and dressed 248 lbs.

The Cotswold and Lincolnshire

Other breeders were not slow in following in Bakewell's footsteps with different breeds, and the Cotswold and Lincolnshire especially, have become the subjects of an equally decided improvement, while the errors of Bakewell were entirely avoided. They possess a rather more desirable robustness, approaching, in some few specimens, almost to coarseness, as compared with the finest Leicesters; but they are more hardy and less liable to disease. They attain as large a size, and yield as great an amount of wool, of about the same value. These breeds scarcely differ more from each other, than do flocks of a similar variety, which have been separately bred for several generations. They are prolific, and when well fed, the ewes will frequently produce two lambs at a birth, for which they provide liberally from their udder till the time for weaning. The weight of the fleece varies from 4 to 8 lbs. per head.

Peculiarity of Long Wool and Its Uses

The striking peculiarity of the long-wools, is in the production of a fleece, which is perfectly adapted, by its length and the absence of the felting property, to the manufacture of worsted stuffs, bombazines, mousseline de laines, &c. This is a branch of manufactures, for which we had little material that was suitable, till the introduction of the long-wools; and its rapid extension in the United States, within the past few years, clearly shows that a large and increasing demand for this kind of wool will continue at remunerating prices. Besides its uses for combing, it is extensively manufactured into blankets, carpeting, and many other fabrics.

Importation of Long-Wools

Several of the Bakewells were imported during the last century; and many flocks, containing some of the best specimens, have been introduced and scattered over every section of the country. The largest of any single importation of the long-wools, was made by Messrs. Corning & Sotham, in 1842, and immediately preceding, and consisted of 70 or 80 choice Cotswolds.

Breeding the Long-Wools

Some information on this subject will be found under the head of *breeding Merinoes*, and *improvement of the Long-Wools*. The ram and ewe should be selected from the best specimens of the breed which is to be perpetuated. There are peculiarities of form or appearance in each, which should be carefully observed. A *violent cross* should never be permitted for the purpose of perpetuation, as suggested under the head of *principles of breeding*, in a previous chapter; such as between those possessing totally opposite properties, as the Merino and long-wools; and there is no conceivable advantage in mixing the middle-wools, South Downs, &c., with either.

Lord Western has long experimented on the blending of the Merino and long-wools, through several generations, without any well-defined results, nor is it believed to be attainable. There is no evenness or integrity of character, either in the animal or fleece, from such mixtures; nor is it possible to foretel the character of progeny from these bastard crosses. The general rule, that like begets like, will not hold true here, for the animal comes large or small, with a long or short fleece, fine or coarse, or intermixed; and this, too, is repeated through numerous generations, when the immediate parents exhibit properties altogether unlike the offspring, and which they derive from some remote ancestry. This practice will do to produce lambs for the butcher, as the consequence of a fresh cross is greater stamina and thrift; and it is found that lambs thus bred attain an early and full development. Thousands of such are annually bred on the banks of the Hudson, Long Island, and around our large cities, and in the worst possible way, as the large, coarse ram is used on the delicate Saxon ewe; yet the lambs thrive and command a good price in the market, and the owner is satisfied to pocket the result. But nothing could be more absurd than to propagate from such progeny for any other purpose than to make early and profitable mutton.

The mixture of breeds of similar character, is attended with the best consequences. Such was the intermingling of the improved Leicesters with the Cotswold and Lincolnshire, by which their former coarseness was removed; and such was the use of the latter with the Leicesters, when they became impotent and almost worthless, from over-refinement in breeding. Good results have followed the mixture of the South and Hampshire Downs. A marked improvement in the Merino in this country, has been claimed by Mr. Jarvis and several others, from the mixture of the various flocks, which for ages had been kept distinct in Spain; and the same result is known to have followed a similar course with the Rambouillet and Saxon flocks.

The Ewe Goes with Young

About five months, varying from 145 to 162 days. Each flock-master will of course determine what is the proper lambing time. For early market, or when there are few sheep, and those well looked after, the lambs may come while the ewes are in the yards, and provision can be made for them, by placing such as are heavy in warm stalls. Both the dam and young thus receive a closer attention than they would in the field; and after a week's housing, in severe weather, the lamb may be turned out into the dry yard, where he will suffer no more, apparently, than the full-grown sheep.

But with large flocks, early lambing is attended with much trouble, and it is generally avoided, by deferring it till the weather has become more settled, and a full bite of grass will afford the dam a plentiful supply of milk. Yet in this case, the young sheep must daily be under the eye of the shepherd, who should see that they are well supplied with food, and especially that they are brought under cover, in severe or stormy weather.

A ram will serve from 20 to 100 ewes in a season, according to his age, health, feed, and management. A South Down or long-woolled lamb, of 7 or 8 months, is sometimes used; and when this is done, he should be well fed, and allowed to run only with a very few ewes. If full-grown rams are turned into a lean pasture to remain with the ewes, not less than four should be put in for every hundred. But if a well-fed ram, in full health and vigor, is kept up, and led out to the ewe as she comes into heat, and allowed to serve her once only, he will suffice for one hundred, without injury to himself or progeny. For this purpose, the ram should be prepared, not by being fat, for this, neither he nor the ewe should ever be; but by being fed with grain for a short time before, and during the continuance of the season. The ewes are more likely to come quickly into heat, and prove prolific, if lightly fed with stimulating food at the time.

It is reasonably enough conjectured, that if procreation, and the first period of gestation, takes place in cold weather, the fœtus will subsequently be fitted for the climate, which rules during the early stages of its existence. If this be so, and it is certainly in accordance with the laws of nature, fine-woolled sheep are most likely to maintain their excellence, by deferring the connection of the male till the commencement of cold weather; and in the Northern states, this is done about the first of December, which brings the yeaning time in the last of April or first of May, when the early grass will afford a large supply and good quality of feed.

Winter Management and Food

Sheep should be brought into winter-quarters soon after the severe frosts occur, as these diminish the feed, and materially impair its nutritious qualities. They ought also to be removed from the grass-lands, before they become permanently softened by the rains, as they will injuriously affect their comfort and health; and allowing them to remain is equally objectionable, from their poaching the sod. If the number be large when brought to the yards, they must be divided into flocks of 50 to 100, according to the size of the yards and sheds. The young and feeble ought to be separated from the others, and the ailing ones placed by themselves; and that no one may suffer from the others, all should be classed as uniformly as possible as to strength. The yards must be dry, well supplied with a trough of fresh water, and with comfortable sheds, to which they can retire when they choose.

Shelters

These, in northern climates, are indispensable to profitable sheep-raising, and in every latitude north of the Gulf of Mexico they would be advantageous. There is policy as well as humanity in the practice. An animal eats much less when thus protected; he is more thrifty, less liable to disease, and his manure is richer and more abundant. The feeding may be done in the open yard in clear weather, and under cover in severe storms. The shelters for sheep are variously constructed, to suit the taste or circumstances of the flock-master. A sheep-barn, built upon a side hill, will afford two floors; one underneath, surrounded by three sides of wall, should open to the south, with sliding or swinging doors to guard against storms; and another may be provided above, if the floors are perfectly tight, with proper gutters to carry off the urine; and sufficient storage for the fodder can be furnished by scaffolds overhead. Or they may be constructed with twelve or fifteen feet posts on level ground, allowing the sheep to occupy the lower part, with the fodder stored above.

In all cases, however, thorough ventilation should be provided, for of the two evils of exposure to cold or too great privation of air, the former is to be preferred. Sheep cannot long endure close confinement without injury. In all ordinary weather, a shed closely boarded on three sides, with a tight roof, is sufficient protection; especially, if the open side is shielded from bleak winds, or leads into a well-enclosed yard. If the floors above are used for storage, they should be made tight, that no hay, chaff, or dust can fall upon the fleece.

Racks or Mangers

These are indispensable to economical feeding. If the hay is fed on the ground, the leaves and seeds, the most valuable part of the fodder, are almost wholly lost; and when wet, the sheep, in their restlessness while feeding, will tread much of it into the mud. To make an economical box or rack, take six light pieces of scantling, say three inches square, one for each corner, and one for the centre of each side. Boards of pine or hemlock, 12 or 15 feet long and 12 or 14 inches wide, may then be nailed on to the bottom of the posts for the sides, which are separated by similar boards at the ends, 2½ feet long. Boards 12 inches wide, raised above the lower ones by a space of 9 to 12 inches, are nailed on the sides and ends, which completes the rack. The edges of the opening should be made perfectly smooth, to prevent chafing or tearing out the wool. The largest dimensions above given, are suitable for the large breeds, and the smallest for the Saxon, and still smaller are proper for their lambs. These should be set on dry ground, or under the sheds, and they can easily be removed wherever necessary.

Some prefer the racks made with slats, or smooth, upright sticks, in the form of the usual horse-rack. There is no objection to this, but it should always be accompanied by a board trough affixed to the bottom, to catch the fine hay which falls in feeding. These may be attached to the side of a building, or used double. A small lamb requires fifteen inches of space, and a large sheep two feet, for quiet, comfortable feeding; and at least this amount of room should be provided around the racks for every sheep.

Troughs

They may be variously constructed. The most economical are made with two boards of any convenient length, ten to twelve inches wide. Nail the lower side of one upon the edge of the other, fastening both into a two or three inch plank, fifteen inches long and a foot wide, notched in its upper edge in the form required.

Food

There is no better food for sheep, than well-ripened, sound, timothy hay; though the clovers, and nearly all the cultivated grasses, may be advantageously fed. Bean and pea straw are valuable, and especially the former, which, if properly cured, they prefer to the best hay; and it is well adapted to the production of wool. All the other straws furnish a good food, and sheep will thrive on them without hay, when fed with roots or grain.

Roots ought to be given them occasionally for a change, and especially to the ewes after lambing, if this occurs before putting them on to fresh pasture. They keep the stomach properly distended, the appetite and general health good; and they render their winter forage nearly equal to their summer feed.

Much grain is not suited to store-sheep. It is too rich, and should be given sparingly except to the lambs, the old ewes, or feeble sheep, or to restore the rams after hard service. For the above purposes, oats are the best; and if any other grain, beans or peas are given, it should be in small quantities. When there is a deficiency of hay and roots, grain may be used with straw.

The flock ought to be so fed as to receive the same amount of nourishment throughout every part of the year. The evenness and value of the fleece depends much upon this. When the amount of nutrition is great, the wool-secreting organs are distended, and the fibre becomes enlarged; when limited, they necessarily contract, and the fibre is small. This produces a want of trueness, which the experienced stapler readily detects, and which he does not fail to estimate against the value of the fleece.

Sheep ought to have a full supply of salt, and if accessible, sulphur, ashes, tar, and clay would frequently be nibbled by them when their stomach required either. Pine or hemlock boughs are a good substitute for tar, and afford a most healthful change in the winter-food of sheep. Entire cleanliness and dryness are also essential to the health of the flock. The smaller sizes of the Saxon may be well sustained on two pounds of hay, but larger sheep will consume from three and a half to four, or even five pounds per day. Sheep, like all other animals when exposed to cold, will consume much more than if well protected, or than during a warmer season.

The Care of the Ewes with Young

Is an important consideration, as the lamb is sometimes the only profit yielded by the flock; for when fodder is high, or wool low, the fleece will barely pay for the food and attention. Pregnant ewes require the same food as at all other times; but caution is necessary to prevent injury or abortion, which is often the result of excessive fat, feebleness, or disease. The first may be remedied by blood-letting and spare diet; and both the last by restored health and generous food. Sudden fright, as from dogs or strange objects; long or severe journeys; great exertions; unwholesome food; blows in the region of the fœtus, and some other causes, produce abortion.

Yeaning

Most flocks are turned into the pasture before yeaning time, and the ewe is then left to nature, which is a good practice, if she is healthy and the weather good. But a larger number of lambs will be reared by a careful oversight of the ewes, and the use of proper precautions. As their time approaches, which may be known by the springing of the udder and the enlargement of the natural parts, they should be put by themselves at night, in a warm stable or with others in the same condition, and well looked after, late and early in the day.

They seldom need any assistance, nor should any be rendered, except in case of wrong presentation, or feebleness in expelling the fœtus. In the former case, the shepherd may apply his thumb and finger, after oiling, to push back the young, and assist in gently turning it till the nose and fore-feet appear and for the latter, only the slightest aid should be rendered, and that to help the throes of the dam.

Management of Lambs

When lambing in the field, only a few should be together, as the young sometimes get changed, and the dams refuse to own them. This difficulty is generally obviated, by holding the ewe till the lamb has sucked two or three times; or they may be shut up together, and the lamb rubbed with a little fine salt. The lamb does not require nourishment for some hours after its birth; but if the dam refuse to lick it as soon as it appears, it must be carefully wiped dry. If the weather be cold and the lamb is dropped in the field, the shepherd should be furnished with large pockets or a well-lined basket, in which it must be placed till the ewe is brought to the shed.

After the first day or two, the udders ought to be completely drained of their milk by the hand, so as to prevent swollen or caked bag. In case of deficiency of milk, the lamb may be supplied from a new milch cow, by means of a sucking-bottle with an air vent, or it may draw a part of its nourishment from another ewe, which can be held while the lamb is sucking.

It is sometimes necessary to substitute a foster-mother, in which case, the ewe may be made to own the lamb, by milking from her udder over the lamb and under his tail, rubbing it on well; or rub the adopted lamb with the entrails and contents of the stomach of the dead lamb, or cover it with the skin. If the ewe proves a bad nurse, or it is desirable to bring the lambs forward rapidly, they may be early taught to eat boiled oats or other grain, cabbage, roots, and tender hay. Lambs should be well fed, as this is important to produce size, constitution, and perfection of form.

The ewes and their young ought to be divided into small flocks, and have a frequent change of pasture. Some careful shepherds adopt the plan of confining their lambs, and allow them to suck two or three times a day; by which they suffer no fatigue, and thrive much faster. But this is troublesome and injurious, as the exercise is essential to the health and constitution of the lamb intended for rearing. It is admissible only when they are wanted for an early market, and by those who rear them for this purpose, it is a common practice.

Castrating and Docking Lambs

After selecting enough of the choicest rams for stock-getters, the castrating may be performed at any time between two and six weeks old, when the lamb is in good health. A cool day should be chosen; or if warm, it must be done early in the morning. The best method is for one person to hold the lamb firmly between his legs, on an inclined plank upon which he rests, while another with a sharp knife cuts off about two-thirds of the lower part of the scrotum. The testicles are then drawn out till the spermatic cord is reached, which is divided by the thumb nail; or it is pulled out and cut with a sharp knife. It is sometimes done by simply opening the scrotum, when the testicles and spermatic cord are jerked out. The wound should then be rinsed with cold water, after which apply lard.

The operation of docking is by many deferred till a late period, from apprehension of too much loss of blood; but if the weather be favorable, and the lamb in good condition, it may be performed at this time with the least trouble and without injury. The tail should be laid upon the plank, the person holding him in the same position as before. With one hand he draws the skin towards the body, while the other person, with a two-inch chisel and mallet, strikes it off at a blow, between the bone joints, leaving it one and a half to two inches long. The skin immediately slips back over the wound and is soon healed. Ewe lambs should be docked closer than the rams. To prevent flies and maggots, and assist in healing, it is well to apply an ointment composed of lard and tar, in the proportions of four pounds of the former to one quart of the latter. This is also a good application for the scrotum. The lambs should be carefully protected from cold and wet till they are perfectly well.

Tagging, or Clatting

Is the removal of such wool as is liable to get fouled, when the sheep are turned on to the fresh pastures, and of course it should be done just before leaving their winter quarters. It is most easily accomplished by placing the animal on a low table, and then holding it as in shearing, till the operation is performed. All the wool near the extremity of the sheath, and the scrotum of the males; from the udder of the ewes; and from below the dock, the inside of the thighs, and the legs of the sheep, should be removed.

Summer Management

As soon as the warm weather approaches, and the grass appears, sheep become restive and impatient for the pasture. This instinct should be repressed till the ground has become thoroughly dry, and the grass has acquired substance. They ought, moreover, to be provided for the change of food, by the daily use of roots for a few days before turning out. It would also check the tendency to excessive purging, which is induced by the first spring feed, if they were housed at night, and fed for the first few days with a little sound, sweet hay. They must be provided with pure water, salt, &c., as in winter, for though they may sometimes do tolerably well without either, yet thrift and freedom from disease are cheaply secured by this slight attention.

Dry, sweet pastures, and such as abound in aromatic and bitter plants, are best suited for sheep-walks. No animal, with the exception of the goat, crops so great a variety of plants. They eat many which are rejected by the horse and the ox, and which are even essential to their own wants. In this respect, they are valuable assistants to the husbandman, as they feed greedily on wild mustard, burdocks, thistles, marsh-mallows, milkweed, and various other offending plants; and the Merino exceeds the more recent breeds in the variety of his selections.

Many prepare artificial pastures for their flocks. This may be done with a number of plants. Winter rye, or wheat sown early in the season, may be fed off in the fall, without injury to the crop; and in the following spring, the rye may be pastured till the stalks shoot up and begin to form a head. This affords an early and nutritious food. Corn may be sown broadcast, or thickly in drills, and either fed off in the fields, or cut and carried to the sheep in their folds. White mustard is a valuable crop for this purpose.

To give sheep sufficient variety, it would be better to divide their range into smaller ones, and change them as often at least as once a week. They seek a favorite resting-place, on a dry, elevated part of the field, which soon becomes soiled. By removing them from this for a few days, rains will cleanse, or the sun dry it, so as again to make it suitable for them. More sheep may be kept, and in better condition, where this practice is adopted, than where they are confined to the same pasture.

Washing Sheep

In most of that portion of the Union north of 40°, the washing is performed from the middle of May till the first of June, according to the season and climate. When the streams are *hard*, which is frequently the case in limestone regions, it is better to do this immediately after an abundant rain, by which the lime derived from the springs is proportionally lessened. The practice of a large majority of our farmers, is to drive their sheep to the washing-ground early in the morning on a warm day, leaving the lambs behind. The sheep are confined on the bank of the stream by a temporary enclosure; from which they are taken, and if not too heavy, are carried into water sufficiently deep to prevent their touching bottom. They are then washed, by gently squeezing the fleece with the hands, after which they are led ashore, and as much of the water pressed out as possible before letting them go, as the great weight retained in the wool frequently staggers and throws them down.

A good practice is to lead the sheep into the water and saturate the fleece, after which they are taken ashore. When they commence *steaming*, they are again led into the water, and washed clean. This insures thorough cleansing, where the water is pure. Others make use of a boat, one end of which rests on a bold shore, and the other is in deep water. The operator stands in the boat and plunges the animal over the side, when the washing is performed. It is sometimes done by sinking a tight hogshead or large box in the water, with heavy weights, in which a man stands, and the sheep are brought or led to him by another person, who walks on a platform reaching from the bank to the hogshead. Either of the last methods obviates the necessity of standing for a long time in water, by which colds, rheumatism, &c. are frequently contracted. In parts of Germany, and sometimes in this country, sheep are forced to swim across a narrow stream several times, by which the fleece is tolerably cleaned, if all the water be pressed out when they get to the land. The yolk being a saponaceous compound, and not an oily matter as is generally supposed, it readily combines with the water and passes out of the wool.

An excellent practice, when streams are not convenient, is to lead a small ripple of soft water into a tub. To this, a little soap is added, after which the sheep are immersed and thoroughly cleansed. Perfect whiteness and purity of the fleece is readily secured afterwards, by throwing over the sheep a jet of water. This practice has a good effect, in preventing or removing cutaneous disorders, and destroying ticks or other vermin.

Many judicious farmers object to washing sheep, from its tendency to produce colds and catarrhal affections, to which sheep are particularly subject; but it cannot well be dispensed with, as the wool is always more saleable, and if carefully done, need not be attended with injury. Warm settled weather, however, is indispensable to washing with safety to the general health of the sheep.

Figure 26 shows a ewe, with lines indicating the usual method of sorting wool; number 1 indicating the refina or picklock; 2 and 3, the second and third qualities.

Shearing

The manner of shearing varies with almost every district; but as this is an art to be acquired under a skilful master, we shall omit particular details on the subject. First clip all the tags and filth, if any remains or has been accumulated after the tagging in the spring; then take off the fleece and spread it with the outside uppermost on a smooth bench or table, and push the wool carefully together, to render it more compact; double the sides over to the centre; throw the clean loose locks into the middle, and roll together from each end. This makes a smooth, dense package, which is secured by passing a stout twine one or more times around the sides and ends. All the wool from the extremities, should be closely sheared and saved by itself, before dismissing the sheep, but not put up with choice fleeces.

Figure 26. Wool Sorting.

If wounds are made, which is sometimes the case with unskilful operators, a mixture of tar and grease ought to be applied. After shearing, such horns and hoofs as are likely to be troublesome, should be sawed and pared.

The *branding*, or *marking*, is essential to distinguish them from other flocks, and this is done on the shoulder, side, or buttock. A brush or marking-iron is used for this purpose, with paint made of lampblack, to which a little spirits of turpentine is first added, and then diluted with linseed or lard oil.

If the weather be cool, and especially, if severe storms occur after washing or shearing, the flock should be housed. If sultry, they should have a cool, shady retreat, where they will be shielded from the flies and the heat. Blisters and permanent injury to the skin and fleece, are frequently the result of such exposure. Shade trees in their pastures, contribute much to the comfort of sheep, when exposed to a blazing sun. A close examination of the skin should be made at shearing, for the detection of disease or vermin.

For remedies, see article diseases.

Smearing or Salving Sheep

Is a custom little practised in this country. For cold, elevated, and bleak exposures, it may be necessary, and it is, therefore, generally adopted in Scotland. The object is, to prevent cutaneous diseases and vermin, and furnish additional warmth and protection to the fleeces of such breeds as are deficient in yolk. It is usually performed in the latter part of October, but is sometimes done immediately after shearing.

The mixture or salve consists of tar and butter or grease, in different proportions; 1 gallon of the former to 12, or sometimes 20 lbs. of the latter; the greater proportion of tar being required for the younger sheep, or for more exposed situations. The grease is melted over the fire, and the tar stirred in, and when sufficiently cool, it is applied to the whole body of the sheep, by carefully parting the wool and rubbing it on the skin with the fingers. The above quantity is sufficient for 30 or 50 sheep, according to their size and the character of the wool.

This application is not required for fine-woolled sheep, whose fleeces are more appropriately protected by a natural secretion of yolk; and it is better to omit it in all cases, where the health and comfort of the animal do not render it absolutely essential. Mr. Stewart, an experienced Scotch shepherd, uses only tallow and train oil, mixed in equal proportions. He asserts, that the improvement in the growth and quality of the wool is at least one-third, and it materially benefits the condition of the sheep.

Weaning

The lambs may be weaned from 3½ to 4 months old. They should be put upon rich, sweet feed, but not too luxuriant; while the dams are turned upon the poorest, and so remote from their young, as to be out of sight and hearing. The ewes ought to be carefully examined after a day or two, and if necessary, the milk removed with the hand. If it continues to accumulate, the ewe may be fed on hay for a few days.

When thoroughly dried off, they should have the best fare, to enable them to recover condition for subsequent breeding and wintering. The fall is a critical period to lose flesh, either for sheep or lambs; and if any are found deficient, they should be at once provided with extra feed and attention. If cold weather overtakes them poor or in ill-health, they will scarcely outlive it; or if by chance they survive, their emaciated carcass, impaired constitution, and scant fleece, will ill repay the food and attention they will have cost.

The Time for Taking Sheep from the Pastures

This must depend on the state of the weather and food. Severe frosts destroy much of the nutriment in the grasses, and they soon after cease to afford adequate nourishment. Long exposure to cold storms, with such food to sustain them, will rapidly reduce their condition. The only safe rule is to transfer them to their winter-quarters the first day they cease to thrive abroad.

Drafting the Flock

For the purpose of ridding it of the supernumeraries, should be done at an earlier day. Such of the wethers as have attained their prime, and those ewes that have passed it, ought to be withdrawn soon after shearing, provided with the best feed, and rapidly fitted for the shambles. If they have been properly pushed on grass, they will be in good flesh by the time they are taken from it; and if not intended for stall-feeding, the sooner they are then disposed of the better.

Stall-Feeding

This will be lost on an ill-shaped, unthrifty beast. The perfection of form and health, and the uniform good condition, which characterize the thrifty one, indicate too plainly to be misunderstood, those which will best repay the care of their owner. The selection of any indifferent animal for stall-fattening, whether cattle or sheep, will inevitably be attended with loss. Such ought to be got rid of when first brought from the pasture, for the most they will bring.

Management of Sheep for the Prairies

When destined for the prairies, sheep ought to commence their journey as early after shearing as possible. They are then disencumbered of their fleece, and do not catch and retain as much dust, as when driven later. Feed is also generally better, and the roads are dry and hard. Young and healthy sheep should be selected, with early lambs; or if the latter are too young, and the distance great, they should be left and the ewes dried off. A large wagon ought to accompany the flock, to carry such as occasionally give out; or they may be disposed of whenever they become enfeebled. With good care, a hardy flock may be driven at the rate of 12 or 14 miles a day. Constant watchfulness is requisite, to keep them healthy and in good plight. One-half the expense of driving, may be saved by the use of well-trained shepherd-dogs.

When arrived at their destination, they must be thoroughly washed, to free them from all dirt, and closely examined as to any diseases they may have contracted, which, if discovered, should be promptly removed. A variety of suitable food and good shelter must be provided, for the autumn, winter, and spring ensuing, and every necessary attention given them. This would be necessary if indigenous to the country; how much more so, when they have just undergone a campaign, to which neither they nor their race have been accustomed!

Sheep cannot be kept on the prairies without much care, artificial food, and proper attention; and from a false system of economy, hitherto attempted by many, losses have occurred from disease and mortality in the flocks, sufficient to have made ample provision for the comfort and security of twice the number lost. More especially do they require proper food and attention, after the first severe frosts set in, which wither and kill the natural grasses. By nibbling at the *fog*, (the frostbitten, dead grass,) they are inevitably subject to constipation, which a bountiful supply of roots, sulphur, &c., is alone sufficient to remove.

Roots, grain, and good hay; straw, or corn-stalks, pea or bean vines, are essential to the preservation of their health and thrift during the winter, everywhere north of 39°. In summer, the natural herbage is sufficient to sustain them in fine condition, till they shall have acquired a denser population of animals, when it will be found necessary to stock their meadows with the best varieties of artificial grasses.

The prairies seem adapted to the usual varieties of sheep introduced into the United States; and of such are the flocks made up, according to the taste or judgment of the owners. Shepherd-dogs are invaluable to the owners of flocks, in those unfenced, illimitable ranges, both as a defence against the small prairie-wolf, which prowls around the sheep, but which are rapidly thinning off by the settlers; and also as assistants to the shepherds in driving and herding their flocks on the open ground.

Diseases of Sheep

The dry and healthful climate, the rolling surface, and the sweet and varied herbage, which generally prevail in the United States, insure perfect health to an originally sound and well-selected flock, unless peculiarly exposed to disease. No country is better suited to sheep, than most of the northern and some of the southern parts of our own. In Europe, and especially in England, where the system of management is necessarily in the highest degree artificial, consisting frequently in early and continued forcing the system, folding on wet, plowed grounds, and the excessive use of that watery food, the Swedes turnip, there are numerous and fatal diseases. Hence the long list which lumbers the pages of foreign writers on sheep.

The most destructive of these are the rot, and epidemics which are scarcely known in America, except by report. The diseases incident to our flocks may generally be considered as casualties, rather than as inbred, or necessarily arising from the quality of food, or from local causes. It may be safely asserted, that with a dry pasture, well stocked with varied and nutritious grasses; a clear, running stream; sufficient shade and protection against severe storms; a constant supply of salt, tar, and sulphur in summer; good hay, and sometimes roots, with ample shelters in winter; young sheep, originally sound and healthy, will seldom or never get diseased on American soil.

The few diseases which it may be necessary here to mention, will be treated in the simplest manner. Remedies of general application, to be administered often by the unskilful and ignorant, must neither be elaborate nor complicated; and, if expensive, the lives of most sheep would be dearly purchased by their application.

A sheep which the owner has reared or purchased at the ordinary price, is the only domestic animal which can die without material loss to its owner. The wool and pelt will, in most instances, repay its cost, while the carcasses of other animals will be worthless except for manure.

The loss of sheep from occasional disease, will leave the farmer's pocket in a very different condition from the loss of an equal value in horses or cattle. Yet humanity, equally with interest, dictates the use of such simple remedies for the removal of suffering and disease, as may be within reach.

Diarrhœa or Scours

When light and not long continued, calls for no remedy. It is a healthful provision of nature for the more rapid expulsion of some offending matter in the system, which, if retained, might lead to disease. It is generally owing to improper food, as bad hay or noxious weeds; to a sudden change, as from dry food to fresh grass; or to an excess, as from overloading the stomach; and sometimes, from cold and wet.

The *remedies* are obvious.

Change to suitable food in the first two cases; enforce abstinence after repletion; and provide warm, dry shelter, with light diet, if owing to the latter causes.

When severe or long continued, a dose of castor oil may be given, and after its operation, give four grains of opium and one ounce chalk, and put them on dry food. Wheat-bran or shorts, and oat-meal or flaxseed ground, are both good for ailing lambs and sheep; as are also ripe oats or wheat, fed in the sheaf, with well cured, sweet hay, and plenty of salt. Fresh boughs of the juniper, or pine and hemlock, help to check the disorder.

Looseness in the larger lambs is prevented by having chalk within their reach; or if they refuse it, administer it in their food. When it happens soon after birth, place it with the ewe in a warm place, and feed the latter with plenty of oats, or other sound grain. If the milk be deficient, give the lamb cow's milk scalded, or let it suck the cow. The tail is sometimes glued on to the buttocks, while the scours continue. Separate it immediately by the use of warm water, and rub the parts with dry loam or clay.

Dysentery

This is a different and frequently a fatal disease, but resembles the former in its general symptoms. It is owing to prolonged diarrhœa, unwholesome or meager food, and other causes. Bleeding and physic should be resorted to, after which give warm, nourishing gruel.

Hoven

For description and remedies, see hoven in cattle.

Braxy

This is manifested by uneasiness, loathing of food, frequent drinking, carrying the head down, drawing the back up, swollen belly, feverish symptoms, and avoidance of the flock. It appears mostly in late autumn and spring, and may be induced by exposure to severe storms, plunging in water when hot, and especially by constipation brought on by feeding on frostbitten, putrid, or indigestible herbage.

Remedies are not often successful, unless promptly applied. Bleed freely, and to effect this, in consequence of the stagnant state of the blood, immersion in a tub of hot water may be necessary. Then give two ounces Epsom salts, dissolved in warm water, with a handful of common salt.

If this is unsuccessful, give a clyster made with a pipe-full of tobacco, boiled for a few minutes in a pint of water. Administer half, and if this is not effectual, follow with the remainder. Then bed the animal in dry straw and cover with blankets, and assist the purgatives with warm gruels, followed by laxative provender till well.—(*Blacklock.*) Thousands of sheep have died on the prairies from braxy, induced by exposure and miserable forage. Entire prevention is secured by warm, dry shelters, and nutritious, digestible food.

Costiveness

This is removed by giving two tablespoonfuls of castor oil every twelve hours, till the difficulty is removed; or give one ounce Epsom salts. This may be assisted by an injection of warm, weak suds and molasses.

Stretches

Sheep sometimes stretch out their noses on the ground, and around their sides, as if in severe pain. This may be caused by an involution of one part of the intestine within another. When owing to this cause, the difficulty is frequently removed by jerking the animal by the hind-legs several times, when the pain disappears.

But it is generally occasioned by *costiveness*, which see above. This may be prevented by using green food, roots, &c., once a week, or by allowing them to browse on the evergreens, pines, hemlock, and firs.

Poison

From laurel and other plants, is cured by pouring a gill of melted lard down the throat, or boil for an hour the twigs of the white ash, and give half to one gill of the strong liquor immediately; to be repeated if not successful.

Inflammation of the Lungs

This is produced by improper exposure to cold and wet. The remedy for slight affections, is warm, dry shelter, and light food. When severe, resort must be had to bleeding and purging freely, then to light bran or linseed mashes.

Rot

This sometimes causes the death of a million of sheep in a single year in England, yet it is a disease almost unknown in this country. Foreign authorities ascribe it entirely to excessive humidity of climate, wet pastures, or too watery food.

The preventives are therefore obvious. After the use of dry food and dry bedding, one of the best is the abundant use of pure salt. In violent attacks, early bleeding, followed by a dose of two ounces Epsom salts, to be repeated if necessary, with a change of diet and location, is all that can be done.

Foot-Rot

This is frequently a prevalent disease among American sheep. It is sometimes spontaneous, but more often produced by contagion. In the former case, it is caused by soft, rich, or moist pastures. A dry gravelly or rocky range, will of course be an effectual preventive when owing to this cause.

The disorder is communicated by the absorbents of the foot coming in contact with the suppuration which has been left on the ground from the diseased part. Absolute safety against this contagion is secured only by a total avoidance of the walks of the infected animals, till repeated rains, or what is better, frosts, have disarmed the virus of its malignity.

Remedies are variously compounded, of blue vitriol, verdigris, tar, spirits of turpentine, alum, saltpetre, salt, lime, copperas, white-lead, antimony, alcohol, urine, vinegar, &c., all of which have proved effectual.

The hoof should first be pared and thoroughly scraped. Then apply a wash made of three parts of blue vitriol, one of verdigris pulverized finely, with scalding (not boiling) vinegar; stirring briskly till it is of the consistence of thin cream, and put it upon the affected part with a paint brush. It is a good preventive, to apply this to the sound feet of the affected animal.

Another remedy is to use spirits of turpentine after scraping; and if the disease is of long standing, add to the turpentine a strong decoction of blue vitriol dissolved in water. The foot should be examined every week, and the remedy repeated till perfect soundness is restored.

A feather dipped in muriatic or nitric acid, and applied to the parts after scraping and cleansing, is a good remedy. When put upon the soles of foot-sore sheep, it hardens the hoofs, and enables them to travel better.

Sheep are sometimes cured by keeping them on a dry surface, and driving over a barn-floor daily, which is well covered with quicklime. It may also be cured by dryness, and repeated washing with soap-suds.

The above ailment should not be confounded with a temporary soreness, or inflammation of the hoof, occasioned by the irritation from the long, rough grasses which abound in low situations, which is removed with the cause; or if it continues, apply white paint or tar, after thorough washing.

Corrosion of the Flesh by Flies or Maggots

May be cured by first removing the vermin; then wash with Castile soap and warm soft water, after which apply white-lead with linseed oil. Tar put on the festering wound corrodes it; but this, or spirits of turpentine placed on the sound parts near it, keep off the flies by their strong effluvia.

If the wound be slight, and the weather moderate, apply a little spirits of turpentine with a strong decoction of elder bark.

Flies on Sheep

May be prevented by smearing with a composition made of two pounds lard or soft grease, one pound sulphur, half pint oil of amber, or oil of tar, or tar alone. A small spoonful is sufficient for a sheep.—*Genesee Farmer.*

Protection from the Gad-Fly

In July, August, and September, in the Northern states, the *gad-fly* (*Œstus ovis*) attacks the nostrils of the sheep, and there deposites its eggs, which, on being hatched, immediately crawl up and make a lodgment in the head. They are frequently repelled by laying a thick coat of tar on the bottom of the troughs, and sprinkling it with salt. The smell of the tar adhering to the nose will drive off the fly. A more effectual remedy is to apply it thoroughly with a brush to the external part of the nose.

If a few furrows of loose earth are turned up in their pastures, the sheep will hold their noses to them, and thus keep off the fly.

The symptoms of grubs in the head, are drooping of the head and ears, discharge of bloody and watery matter from the nostrils, and loss of strength in the limbs.

If worms have made a lodgment, take half a pound of good Scotch snuff, and two quarts boiling water: stir, and let it stand till cold. Inject about a tablespoonful of this liquid and sediment up each nostril, with a syringe. Repeat this three or four times at intervals, from the middle of October till January: the grubs are then small, and will not have injured the sheep. The efficacy of the snuff will be increased, by adding half an ounce assafœtida, pounded in a little water. The effect on the sheep is immediate prostration and apparent death, but they will soon recover. A decoction of tobacco will afford a substitute for snuff.—*N. Eng. Far.*

Blacklock's remedy is, to half fill the bowl of a pipe with tobacco, light it, and then hold the sheep, while a person inserts the stem some distance into the nostril, and blows a few whiffs into the nose. The operation is then repeated with the other nostril.

Swollen Mouth

Is sometimes fatal. It is said to be cured by daubing the lips and mouth plentifully with tar.—*Albany Cultivator.*

Foul Noses

Dip a small swab into tar, then roll in salt. Put some on the nose, and compel the sheep to swallow a small quantity.—*American Far.*

A disease indicated by drooping, running at the eyes, weakness in the back and loins, inability to use the hind legs, was removed by turning the sheep into a pasture containing lobelia, (*Indian tobacco.*) Dried lobelia was also given, and produced the same effect.—*Cultivator.*

Scab

This loathsome disease, to which fine-woolled sheep are particularly liable, is caused, like itch in the human subject, by a small insect, a species of the *acari*. It is first manifest by the rubbing of the sheep, and soon after by one or more tufts of wool, which is loosened at the roots. On feeling the skin, a hard, dry tumor is perceptible. To prevent contagion, remove the infected sheep to a separate pasture or yard as soon as discovered.

Remedies.—The Spanish shepherds dissolve a little salt in their mouth, and drop it upon the infected part.

When the tumor has become enlarged, the wool should be removed closely to the skin, the scab scraped with a curry-comb, then wash with strong soap-suds or ley, and afterwards rub thoroughly with sulphur or brimstone, mixed with lard or grease.

An effectual remedy is prepared by taking one pound of tobacco, which add to 12 quarts ley from wood ashes of sufficient strength for washing, and four quarts urine; to this add another mixture of a gill high-wines; ¼ oz. camphor; ¼ oz. Spanish brown, and ½ gill spirits of turpentine. A small quantity of this applied to the sore will never fail.

Immediately after shearing, scab may readily be cured by immersing the sheep, (excepting the head,) in a strong decoction of tobacco liquor, adding a gill of spirits of turpentine for the first, and making a slight addition of fresh liquid for each sheep, enough to keep up the strength of the tobacco and turpentine, and taking care to rub the affected part thoroughly. For lambs, this liquor should be diluted, but yet left strong enough to kill ticks in one or two minutes, which may be ascertained by experiment.

Scab is also removed by using a composition of one pound plug tobacco to three gallons of water, with lime-water and oil of vitriol added; or a decoction of hellebore with vinegar, sulphur, and spirits of turpentine.—(*H. D. Grove.*)

Scab is propagated more by using the same rubbing posts, than by contact with each other. Sheep in low condition are more subject to it than others.

Ticks and Lice

Sometimes infest sheep. Good feeding and shelter is a partial preventive, but when they have made their lodgment, they must be dipped in a decoction of tobacco water. The most effectual time for their destruction, is a few days after shearing, when they will have left the naked bodies of the old ewes, to hide in the fleeces of the lambs. The dipping in tobacco water, with the addition of a small quantity of turpentine, is an effectual remedy.

After dipping the sheep or lambs, the liquor should be pressed out from the wool, upon an inclined plane, so arranged as again to run into the vessel.

Pelt-Rot

Will be recognised as one of the staple diseases of our *native sheep*, described on page 89. The wool in this case falls off, leaving the sheep partially or almost wholly naked; but this is not accompanied with soreness or apparent disease.

The animal must be provided with a warm stall and generous feed, and the naked skin should be anointed with tar and grease. The preventive is good keeping and shelter.

Staggers or Sturdy, and Water in the Head

Sometimes affect sheep, but more especially lambs under a year old. The first is caused by the *hydatid*. It is considered as an almost incurable disorder, but is sometimes removed by trepanning.

Chancellor Livingston carefully supplied two thus attacked, with food for three months, when nature effected a cure. Removal to dry lands and purging, is a good precaution when they are first taken.

An English lad lately cured one which had been given up, by boring with a gimlet into the soft place on the head, when the water rushed out, and the sheep immediately followed the others to the pasture. A correspondent of the Albany Cultivator asserts, that ½ a pint of melted lard poured down the throat, will cure blind staggers in ten minutes.

Abortion

Occurs sometimes, and is usually caused by excessive fright or exertion, and sometimes by severe exposure and poor feed. It is seldom fatal, except to the lamb.

The Uterus

Is occasionally protruded after lambing. It should be immediately returned, first washing it in warm milk and water, if any dirt adheres to it. For this, the hand only should be used. After rubbing it with lard or oil, hold up the hind legs, and gently replace the protruded parts, then keep the ewe quiet till fully recovered.

For Garget, or Caked Bag

Keep the bag thoroughly drained of milk, for which purpose the lamb is the most efficient. If it is lost, another may be temporarily substituted. Purge freely with Epsom salts, and wash the udder repeatedly with very warm water. If matter forms, it should be opened with the lancet.

Bleeding

"Nothing tends so much to the recovery of an animal from a disease in which bleeding is required, as the rapid flow of the blood from a large orifice. Little *impression* can be made on an acute disease by the slow removal of even a large quantity of blood, as the organs have time to accommodate themselves to the loss, which might, for any good it will do, as well be dispensed with. Either bleed rapidly, or not at all. The nearer the commencement of an ailment, in which you employ bleeding, the operation is resorted to, the greater the chance of its doing good.

Bleeding by nicking the under surface of the tail, does very well, where no great deal of blood is required, but it is not to be thought of if the veins of the face or neck can possibly be opened. These are to be taken in preference to a vein on the leg, as they are much more readily got at. The facial vein commences by small branches on the side of the face, and runs downwards and backwards to the base of the jaw, where it may be felt within two inches of the angle, or opposite the middle grinding tooth. It is here that the orifice must be made: the thumb of the left hand being held against the vein, so as to prevent the flow of blood towards the heart, will make it *rise*.

Some prefer opening the jugular vein, which commences behind the eye and runs down the side of the neck. This vessel is, however, more difficult to open than the former, being better covered with wool, and not so easily exposed or made to swell. *Stringing* is the mode commonly resorted to for this end; that is to say, a cord is drawn tightly round the neck close to the shoulder, so as to stop the circulation through the vein, and render it perceptible to the finger.

A lancet is the instrument generally used in bleeding, though a well-pointed penknife will do at a pinch. The opening must always be made obliquely; but before attempting this, the animal must be secured, by placing it between the operator's legs, with its croup against a wall. The selected vein is then fixed by the fingers of the operator's left hand, so as to prevent its rolling or slipping before the lancet. Having fairly entered the vein, the point of the instrument must be elevated at the same time that it is pushed a little forward, by which motion it will be lifted from or cut its way out of the vein.

A prescribed quantity of blood should never be drawn, for the simple reason that this can never be precisely stated. If the symptoms are urgent, as in all likelihood they will be, your best plan is not to stop the flow of blood till the animal fall or is about to fall. When this occurs, run a pin through the edges of the orifice, and finish by twisting round it a lock of wool."—(*Blacklock.*)

Large Cuts Are Healed

By first sewing and then covering with salve. Smaller ones may be secured with an adhesive plaster or bandage.

To Protect Lambs from Wolves and Foxes

Smear the neck plentifully with a mixture of tar and sulphur. Bells are also said to guard the flock, as both are excessively wary, and have a great dislike to anything artificial. Large dogs will keep them at bay. A better remedy is to kill the marauders, which may be done by inserting strychnine in fresh meat and leaving it in their haunts.

Chapter 5

THE HORSE

In nearly all ages and countries, the horse has been the devoted servant, and the object of the pride and affection of man. Among the semi-civilized Tartars of Middle and Northern Asia; the aborigines of our remote Western prairies, reaching even beyond the Rocky Mountains, and many other rude nations, his flesh is used for food. Most of the tribes among the former use the milk for domestic purposes, and especially when fermented and changed to an unpleasantly sour and intoxicating beverage. But throughout the civilized world, with some slight exceptions, the horse is useful only for his labor. For this purpose he is pre-eminently fitted by his compact, closely-knit frame; his sinewy, muscular limbs; his easy, rapid stride; his general form, and entire structure and habits.

He is found in his wild condition in Central Asia, Siberia, and the interior of Africa; and for 300 years he has been turned loose to follow his native instincts on the illimitable pampas of South America, and the wide-spread prairies of Mexico and California. In all these regions he closely resembles the medium varieties of the domesticated horse; but as the natural result of his freedom, he possesses more fire and spirit than any other, except the blood-horse.

Arabia is generally claimed as the original native locality of the horse, and as the only source from which he is to be derived in the requisite perfection for the highest improvement of the race. But Strabo, who wrote more than 1,800 years ago, asserts that the horse did not then flourish in Arabia, and it was not till some centuries later that he attained any decided superiority there. Great attention, however, has been paid in that country, since the era of Mahomet, to breeding a light, agile, and enduring frame; intelligence and tractability of character; and the perpetuation of these qualities by the most scrupulous regard for the purity of blood.

This is equally true of the Barb or pure-bred horse of Morocco, and those of the northern coast of Africa, in Egypt, among the Turks, and indeed wherever the followers of the Prophet are to be found. It is unquestionable, that the influence of the Eastern blood among the choicest animals of modern Europe, has been followed by great improvements in racing stock. Yet it is equally certain, that the race-horse, both of England and the United States, has accomplished what has never been demonstrated as within the ability of their progenitors; and on repeated trials with the Eastern horses, he has shown himself confessedly their superior in speed, strength, and endurance.

In 1825, two English horses ran against the two fleetest Cossacks which could be found throughout the entire region of their best blood, and in a continued race of 47 miles, the European took the stakes; Sharper, the most successful, performing the distance in 2 hours and 48 minutes. About the same time, Recruit, an English horse of moderate reputation, easily beat Pyramus, the best Arabian on the Bengal side of India.

The Leeds, the Darley, and the Godolphin Arabian; the Lister and D'Arcey's White Turk, and other noted Eastern horses, would not compare in performance with many of their descendants. But these, with some other choice Arabians, on the best mares, and with every advantage for obtaining celebrity, have succeeded in establishing a fame as just as it has been enduring. Yet it must at the same time be remembered, that of the innumerable other pure-bred horses which have been tried in Europe, a few only have rescued their names from oblivion.

The experience of Eastern blood in this country, in comparison with the best English, is decidedly in favor of the latter. We have had one horse of unsurpassed excellence, which a fortunate accident threw upon our shores a short time previous to 1770. This was the white Barb *Ranger*, which was presented by the Emperor of Morocco, as the choice of his stud, to an English naval officer for some distinguished service. On his route homeward, the animal was set on shore for exercise at an intermediate port, where in his gambols he broke three of his legs, and thinking him worthless, his owner gave him to the commander of a New England merchantman, then present. He was readily accepted, and placed in slings on board of his vessel, and recovered.

This animal stood for many years, in the eastern part of Connecticut; and on their good mares, produced a numerous progeny of unrivalled cavalry horses, which rendered invaluable services in the troop commanded by that consummate partisan, Captain (afterwards General) Lee, of the Revolution. It is said the favorite white field-horse of General Washington was of the same stock. He was afterwards sold to Captain Lindsey, as a special favor, and taken, to Virginia, where he produced some good racers.

Bussorah, a small sorrel horse, brought into this country in 1819, from the head of the Persian Gulf, got many choice roadsters, though no racers of celebrity.

The Narraganset pacers, a race belonging to our Northern states, but for many years almost extinct, possessed for a long time an unrivalled reputation for spirit, endurance, and easy rapid motion under the saddle. They are said to have originated from a Spanish horse, many of which are pure descendants of the Barb.

As an offset to these isolated examples of success in this country, we have numerous instances of the importation of the best Orientals, which have been extensively used on some of our superior mares, without any marked effect. We shall refer to three prominent importations only.

The first consisted of two choice Arabians, or Barbs, selected in Tunis by General Eaton, and sent to his estate in Massachusetts. The second was a present of four choice Barbs, from the Emperor of Morocco to our government, in 1830; and the third consisted of two Arabians, sent by the Imaum of Muscat, near the Persian Gulf, to our government in 1840. These were all claimed to be, and no doubt were, of the pure Kochlani, the unadulterated line royal; yet none have earned any distinguished reputation, either by their own performances or those of their descendants.

It is to England we are mainly indebted for the great improvement in our blood, road, and farm horses. A numerous race of fine horses was reared on that island, long previous to its authentic history; for, in his first invasion Julius Caesar took many of them to Rome, where they immediately became great favorites, although she had already plundered every region of some of their best breeds.

What might have been the particular merit of the English horse at the time of the Norman invasion, is not known, but it is certain that the Saxon cavalry under Harold were speedily overpowered by William, at the battle of Hastings, which at once secured the throne to the Conqueror. History first informs us of the improvement of British horses, by importations from abroad during this reign, which consisted of a number of Spanish stallions. These were supposed to be strongly imbued with the Arabian blood, which had been brought over to that country by the Moors, who had founded the Saracenic empire in the Peninsula, three centuries before. More than a century later, John made some importations from Flanders, to give weight and substance to their draught and cavalry horses. The improvement of their various breeds was afterwards pursued, with more or less judgment and zeal, by other British monarchs, till they reached their highest excellence during the middle of the last century.

Flying Childers, Eclipse, Highflyer, and others on the course, have probably exceeded in speed anything ever before accomplished; while the draught-horse, the roadster, the hackney, the cavalry-horse, and the hunter, attained a merit at that time which some judicious authorities claim has not been since increased. It is even asserted, that some of the more serviceable breeds have been seriously injured by too great an infusion of the blood; while the almost universal absence of long heats on the turf has tended to the improvement of speed rather than bottom in the race-horse.

The improvement of the horse in this country has not been a matter of record or history, till within a comparatively recent period. But it has silently, and with no little rapidity, been going forward for more than a century, till we have obtained a race of animals, throughout the Eastern and Middle states at least, which probably equal those of any other country for adaptedness to draught, the road, and the saddle. This improvement has been mainly brought about by the importation of some of the *best and stoutest of the English blood.* In breeding from these for purposes of utility, particular reference has been paid to strength, enduringness, and speed. No horses surpass our best four-mile bloods; none equal our trotters; and though much inequality exists in those bred for various other uses, yet for profitable service, it is believed, no equal number of animals elsewhere can exceed those in the region above indicated.

It would be a superfluous task to attempt enumerating all the imported horses that have contributed to this improvement. Each good animal has done something. But among the earlier horses which may be named with distinction, as having effected much for our useful beasts, are Lath, Wildair, Slender, Sourkrout, Tallyho, Figure, Bay Richmond, Expedition, Baronet, and a host of others.

Pre-eminent among these, was imported Messenger. He was foaled in 1780, imported in 1788, and died in 1808. He stood in different places in New Jersey, and in Dutchess, Westchester, and Queens counties in New York; and upon the mares derived from the foregoing and other good horses, he got a numerous progeny of illustrious descendants. Of these we may name those capital stallions, Potomac, Hamlintonian, Bay Figure, Engineer, Mambrino, Tippoo Saib, Columbus, Gunn's, and Bushe's Messenger, and many others, which were extensively disseminated over the Northern and Middle states; and he has the credit of imparting a large share of his merits to that nonpareil of horses, his grandson, American Eclipse. His posterity were so numerous and widely spread, that it may be safely asserted, that of the best horses bred in the above states, scarcely one can now be found which does not trace one or more crosses to this distinguished sire. His success in producing roadsters, besides his blood qualities of speed and endurance, consisted in his great strength, and the peculiar formation of his limbs, large forehand and deep quarters, in which he excelled any other of the imported bloods.

As an illustration of what may be accomplished by judicious breeding with the present materials in our hands, we mention one family of the American roadster, which is strongly tinctured with blood, and which has attained an enviable notoriety among the choicest of the Northern horses. They are derived from the *Morgan horse* of Vermont, that was foaled in Springfield, Mass., in 1793. He was got by True Britton, supposed to have been bred by Gen. Delancey of New York, and got by imported Wildair, (or one of his sons,) a horse of such distinguished excellence, as to have been re-exported to England, for the benefit of his stock.

The Morgan horse stood in Vermont from 1795 till his death, at an advanced age. From him and the choice mares of Vermont, descended many excellent colts; and his merits were inherited in an eminent degree by three of his sons, which stood in the same state and continued the career of improvement commenced by the sire. The result has been the production of a family of roadsters, of much similarity of appearance[3] and uniformity of character, unsurpassed by any others for serviceable qualities.

They are of medium size, from 13½ to 15 hands high; with a well-formed head and neck; high withers; deep chest; round body; short back; long quarters; broad flat legs; moderately small feet; long wavy mane and tail; presenting altogether the beau ideal of the road horse. They are spirited, docile, hardy, and easily kept. They have an easy, rapid trot, and glide along with a good load, without clatter or apparent effort, at the rate of ten or twelve miles an hour.

This family of horses has not of course been bred long enough within themselves, to have attained to the eminence of a distinct breed. They are mentioned, merely as a type of what the serviceable roadster ought to be, and what he may become by the use of the proper instruments for breeding. And if the materials already in our hands are intelligently and perseveringly used, we can produce all we require of horse-flesh.

---

[3] Many of the Morgan horses have the steep rump and heavy breast and neck, which indicate a Norman cross on the side of their dams. These have been largely imparted through the French horse in the adjoining Canadian settlements, but none of these are said to have characterized the founder of the race.

Besides our unsurpassed blood-horses, we have others derived from various sources, and especially from the different English breeds, all of which are variously compounded, with the first and with each other. On our northeastern frontier, the *Canadian* prevails, a bastard but not degenerate race, made up of the French Norman and the English or American. At the extreme South and West, we have the *horse of Spanish origin*, obtained in his domestic state in Florida and Louisiana; and from another branch of the Spanish, are descended the wild horses of Mexico and the more northern prairies. These are diversified in character, and generally possess medium size and merit.

The *Conestoga*, a heavy roadster, is principally reared in Pennsylvania, and is used for the team and truck. He is an amalgamation of several breeds, but probably owes a share of his character to the Flemish horse, for which there was a decided partiality among the numerous German emigrants of that state.

Figure 27. Norman Horse.

Several varieties of ponies are to be found in different sections, but principally among the French, the half-breed, and the Indians upon the frontiers, who have bred a stunted race from the Canadian or wild-horse, and such others as could survive the hard usage and scanty winter food afforded by nature and their rude husbandry. Many of these have considerable beauty and symmetry, and are fleet, hardy, and spirited.

The modern Norman, or mixture of the old French Norman draught-horse, (heavy-framed, big-limbed, but stout and hardy,) and the Andalusian, a descendant of the Moorish barbs, has been introduced within a few years, and will unquestionably become a very popular horse for many purposes. He exhibits the qualities of both ancestry in the proper proportions for farm service. He has a thick head; lively, prick ears; short, heavy neck; large breast and shoulders; strong limbs; well-knit back; large quarters, with much wavy mane, tail, and fetlock. Like his French progenitor, he frequently stands low in the withers, which enables him to throw great weight into the collar; and the diminished, flattened leg, the wind and game derived from his Moorish blood, give him much of the capacity and endurance of the thorough-bred.

The English cart-horse has for a long time made up some of the best, heavy dray-horses in the country, and late importations have refreshed the breed with additional choice specimens. The Cleveland bay has been introduced of late, and promises good carriage-horses from our well-spread, sizeable mares. The Norfolk trotter Belfounder was imported many years since, and with our high-bred mares, has produced many choice roadsters and trotters.

The remainder of our horse-flesh deserving of any notice, is chiefly composed of such as are superior in point of blood. The improvement in the American horse, from this source, is conspicuous and decided. Judicious breeders still look for qualities in the descendants, which they sought for in their imported sires, and the infusion of some of the stoutest of the blood is rapidly gaining an ascendency in the general stock. Our intelligent agriculturists should look to this subject closely, and not permit this system to proceed to an extent that may be prejudicial to their value as draught-horses, as has been done in some portions of England and our Southern states.

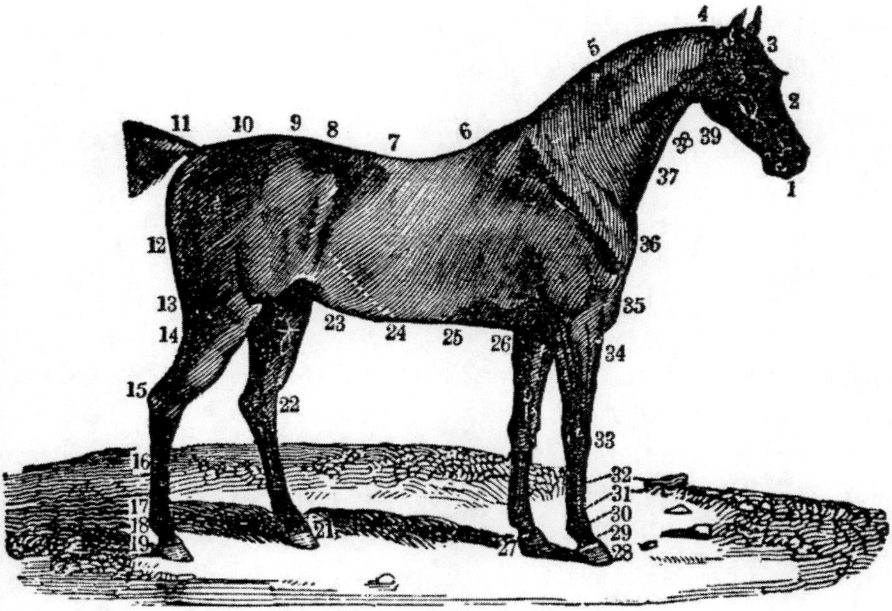

Figure 28. Points of a Horse. 1. Muzzle.—2. Race.—3. Forehead.—4. Poll.—5. Crest.—6. Withers.—7. Back.—8. Loins.—9. Hip.—10. Croup.—11. Dock.—12. Quarter.—13. Thigh, or Gaskin.—14. Ham-string.—15. Point of the Hock.—16. Cannon.—17. Fetlock.—18. Large Pastern.—19. Small Pastern.—20. Hoof.—21. Coronet.—22. Ham, or Hock.—23. Sheath.—24. Flank.—25. Girth.—26. Elbow.—27. Heel.—28. Hoof.—29. Small Pastern.—30. Large Pastern.—31. Fetlock.—32. Cannon.—33. Knee.—34. Arm.—35. Breast, or Bosom.—36. Point of the Shoulder.—37. Windpipe.—38. Gullet.—39. Jowl.

There is no danger from excess of blood, if it be of the right kind; but it is seldom found combining that fulness and stoutness, and that docility and tractableness of disposition, which are essential to the gig-horse or the horse of all work. Yorke says truly, that "the road-horse may possess different degrees of blood, according to the nature of the country and the work required of him. [He might have added with propriety, and according to the character of the blood.] His legs will be too slender; his feet too small; his stride too long, and he will rarely be able to trot. Three parts, or half, and for the horse of all work, even less than that, will make a good and useful animal." For the saddle only, the high-bred, if not disposed to be vicious, is never objectionable to an enterprising and accomplished rider. His long elastic pasterns, giving easy, flexible motions; his quick and almost electrical obedience when under thorough discipline; his habitual canter and high spirit, always commend him for this purpose.

Some of the Prominent External Points of a Fine Saddle or Gig Horse

Are, a moderately small head, free from fleshiness; fine muzzle and expansive nostrils; broad at the throat and wide between the eyes, which denotes intelligence and courage; a dished face indicates high breeding, and sometimes viciousness; a convex or Roman nose frequently betokens the reverse; the ears rather long, yet so finely formed as to appear small, and playing quickly like those of a deer; the eyes clear, full, and confident, with a steady forward look. Glancing them backward or askance with a sinister expression, and with none or only a slight movement of the head, is indicative of a mischievous temper.

The neck should be handsomely arched, and fine at the junction with the head, while the lower extremity must be full and muscular, and well expanded at the breast and shoulders. The latter ought to be high and run well back; the withers strong, firmly knit, and smooth; the breast neither too prominent nor retreating, too wide nor too narrow, and supported by a pair of straight fore-legs, standing well apart. The chest should be deep, and the girth large; the body full, and not drawn up too much in the flank; the back short, and the hips gathered well towards the withers; the loins wide and rising above the spine; the ribs springing nearly at right angles from the back, giving roundness to the body. The hips ought to be long to the root of the tail, and the latter may approach to near the line of the back, which is a mark of good breeding.

Both the thigh and hock should be large and muscular; and between the hock or knee and pastern, the legs should be broad, flat, and short; the hind legs properly bent, and all well placed under the body; the pasterns of moderate length, and standing slightly oblique; the hoof hard, smooth, round before, and wide at the heel; the frog large and sound; and the sole firm and concave. A white hoof is generally tender, easy to fracture and to lame, and difficult to hold a shoe.

The *draught-horse* ought to differ from the foregoing, in possessing a heavier and shorter neck; a wider and stouter breast, and low withers, so as to throw the utmost weight into the collar; a heavier body and quarters; larger legs and feet; and more upright shoulders and pasterns.

Considerations Which Affect the Value of the Horse

The color is not material, provided it be not pied or mealy. No better color for horses can be found than the dark bay or brown, with black mane, tail, and legs. But most of the other colors are frequently found with the best horses.

Hard-mouthed horses, when accompanied with great spirit, are objectionable, as they require peculiar bitting and the utmost vigilance.

The paces and action of a horse are important, for if good they give a much greater capacity for performance. Some of these depend on form and structure, and are unchangeable; others are the result of breaking. All horses should be taught to walk fast, as it is their easiest and most economical pace, and it will help them over a great deal of ground in a day, even with a heavy load, and with comparatively little effort. A horse that steps short and digs his toes into the ground, is worthless as a traveller, and suited only to a ferry-boat or bark-mill.

It is important that a horse be good-tempered. If inclined to viciousness, he should be gently yet firmly managed when it is first apparent. A resort to great severity will be justified, if necessary to conquer him; for if once allowed to become a habit, it will be difficult to cure him. Grooms and mischievous stable-boys, frequently do much injury by their idle tricks with horses; and when detected, they should be discharged at once.

Some horses are nervous, easily excited, and start at every unusual noise or object. Others are restive and fretful, and ever anxious to be on the move. Kindness, and firm, yet mild treatment, by which their motions and will are at all times controlled, and their confidence secured, are the only remedies.

Others are inclined to sluggishness. These should have stimulating food, and never be overloaded or overworked, and then kept well to their paces. Whatever they are capable of performing, can be got from them in this way only. Habit has great influence with animals, as with man; and when within the compass of his ability, he may be habituated to any reasonable physical exertion.

Breeding

Agreeably to the general principles before enumerated, such animals should be selected, as most eminently possess those points which it is desired to propagate, and these, they should not only exhibit in themselves, but should inherit as far as possible, from a long line of ancestry. For the perpetuation of particular points in progeny, it would be safer to rely on the latter quality than the former.

The selection of a mare, relatively larger than the horse, is an important rule in breeding, and it is believed that much of the success of Arabian and other Eastern horses as stock-getters, has resulted from the application of this principle. They possess valuable traits, but condensed within too small a compass. When such an animal is put to a well-bred, larger mare, the fœtus has abundance of room and nourishment to develop and perfect the circumscribed outlines of the male parent, and acquire for itself increased volume and character.

The horse ought not to be less than four or five, and the mare one year older, before being put to breeding. It would be still better to defer it for two or three years, or till the frame is fully matured.

The Gestation of the Mare

Sometimes varies from 44 to 56 weeks, but she usually goes with young from 47 to 50; and it is advisable she should take the horse at a time which will ensure the foaling when the weather is settled, and there is a fresh growth of grass. She will be the better for light working till near the time of foaling, if well, but not too abundantly fed. In a few days after this, she may resume moderate labor; and if not in the way or troublesome, the foal may run with her; but if she is exposed to heating, it should be confined till she cools, as suckling then is decidedly injurious to it.

The mare is in danger of slinking her foal from blows and over-exertion, the use of smutty grain, foul hay, or offensive objects or smell; and when this has once occurred, which happens usually in the fourth or fifth month, she should afterwards be generously fed at that period, and only moderately worked, to prevent a recurrence of the casualty. When liable to slinking, the mare should be removed from others in foal, lest a peculiar sympathy should excite an epidemic.

The Mare Comes in Heat

From nine to eleven days after foaling, when she should be put to the horse, if it be desirable to have a colt the following season. She comes round at intervals of about nine days.

Management of the Colt

The colt may be weaned when five to seven months old; and preparatory to this, while with the mare, may be taught to feed on fine hay, meal, or oats. When taken away, he should be confined beyond a hearing distance of the dam, and plentifully supplied with rowen or aftermath hay, crushed oats, or wheat shorts. It is economy to provide a warm shelter through the inclement season for all animals, and especially for colts, which, with all other young, should have an abundance of nutritious food. They will thus grow evenly and rapidly, and attain a size and stamina at two years old, they would not otherwise have acquired at three.

Castrating

The colt should be altered at about one year, but if thin in the neck and light before, the operation may be deferred to such time as the requisite development is secured. Few of the French diligence and farm horses, and scarcely any of the Oriental, are ever castrated. They are thought to be more hardy and enduring; but the slight advantage they may possibly possess in this respect, would hardly compensate for the trouble and inconvenience too frequently arising from their management.

The operation should be performed late in the spring or early in autumn, while the weather is mild. If in high condition, the animal must first be bled and physicked. If large and fractious, he must be cast. Some back him into the angle of a worm fence, where he is firmly held by the head with a bridle, and the operator accomplishes the object without any trouble or material restiveness from the animal while standing. The scrotum should be opened on both sides, and the testicles cut, or rather the cord scraped off, which prevents much bleeding. The wound may be dressed with a little lard; then turn him loose in a pasture which has a shelter from sun, wind, or rain. Another method of castrating is by *torsion*, or twisting.

Docking

Is practised by many, but merely to gratify an absurd and cruel caprice, without a single advantage, and the animal is better in every respect with the tail unmutilated. If done at all, it should be when young, and with a single stroke of the knife, or chisel and mallet; and if the weather be favorable, no further attention is necessary.

Nicking

This inhuman custom is now getting unfashionable, and we omit any description of it.

Breaking

While feeding in the stable, the colt should be gently treated, and accustomed to the halter and bit, which prepares him for breaking. If permitted to run with the others while at work, he becomes familiarized to it, and when harnessed by the side of some of his well-trained mates, he seems to consider his discipline rather a privilege than a task. The colt may be taken in hand for breaking at three years of age, and thoroughly broken to light work at four, but should not be put to hard service till six or eight. A due regard to humanity and sound judgment, in thus limiting the burden in his early years, would save much disease and suffering to the animal, and profit to the owner, by his unimpaired strength and prolonged life. The annual loss from neglecting this precaution is enormous, which might be entirely avoided by less eagerness to grasp the substance, while as yet the shadow only is within reach. Many animals are thus broken down at twelve, and are in their dotage at fifteen, while others of good constitution, if well treated, perform hard service till thirty.

Longevity of the Horse

Mr. Percival mentions one that died at 62. Mr. Mauran, of New York, has a fine gig and saddle horse, now in his 45th year, sound, spirited, and playful as a kitten. He is of a dark brown, with a tanned nose. We never yet saw a horse with a buff or bear muzzle, that had not great endurance.

American Eclipse was successfully covering mares in Kentucky at the age of 32, the result of late and light service till his sinews became fully matured. We have frequently seen a large, compact, flea-bitten horse at work, dragging a heavy load in a single cart, which was formerly used as one of Governor Maitland's coach-horses, and though upwards of 30, he was apparently as sound and vigorous as an overtasked colt of seven or eight.

Feeding

The vigor and duration of the horse depend much on proper feeding. Like the cow and sheep, he may be made to subsist on animal food, fish, and almost every species of nutritious vegetable. But his natural and proper aliment is the grasses, grain, and roots. In the middle and northern sections of this country, his dry forage is almost invariably good meadow-hay, generally timothy, which is the richest of the cultivated grasses. At the South, this is often supplied by the blades of Indian corn. But in all the states, a great variety of the grasses and clover are used.

When put to hard labor, grain ought always to accompany hay in some form. Of the different kinds of grain, oats are peculiarly the horse's food, and they are always safe, digestible, and nutritive. Barley is the best substitute for it. Wheat and Indian corn are sometimes given, but both are unsuitable; the first is too concentrated, and the last too heating. They ought to be sparingly used, and only when ground and mixed with chaff. The offal of wheat is never objectionable.

Grain is always more advantageously fed when ground or crushed, and wet some time previous to eating; and it is still better when cooked. On both sides of the Mediterranean, in the Barbary States, in Spain, France, and Italy, much of the food is given in small baked cakes, and the saving in this way is much greater than the expense of preparing it.

When confined to dry food, roots or apples fed once a day are always beneficial. They keep the bowels open, the appetite and general health good, and contribute largely to the nutriment of the animal. Carrots are the best of the roots, as besides giving muscle and working power, they, more than any other, improve the wind and prevent all tendency to heaves. They have even been found effectual in removing an obstinate cough. By many of the keepers of livery stables, they are always used, for which purpose they command the same price as oats. Potatoes, parsnips, beets, and Swedes turnips, in the order mentioned, are next to be preferred. Potatoes are improved by cooking.

Mixtures of food are best, as of cut hay, meal, and roots. Old horses, or such as are put to hard labor, will do much better if their food be given in the form easiest of digestion. No inconsiderable part of the vital power is exhausted by the digestion of dry, raw food.

Horses ought to be fed, and if possible, exercised or worked regularly, but never on a full stomach. This is a frequent cause of disease, and especially of broken wind. If their food is given at the proper time, and the horse be allowed to finish it at once, without expecting more, he will lie down quietly and digest it. This will be much more refreshing to him, than to stand at the rack or trough, nibbling continually at his hay or oats. What remains after he has done feeding, should be at once withdrawn.

They should have water in summer three times, and in winter twice a day. Soft or running water is much the best. While working, and they are not too warm, they may have it as often as they desire. Neither should they be fed when heated, as the stomach is then fatigued and slightly inflamed, and is not prepared for digestion till the animal is again cool. Salt should always be within reach, and we have found an occasional handful of clean wood-ashes, a preventive of disease and an assistance to the bowels and appetite.

DISEASES

The list is long and fearful, and even the brief one subjoined, will be found sufficiently great to inculcate the utmost caution in their management. The horse in his natural condition is subject to few ailments. It is only in his intensely artificial state, and when made the slave of man, that he becomes a prey to disease in almost every shape. A careful and judicious attention to his diet, water, exercise, stable, and general management, will prevent many of those to which he is subject.

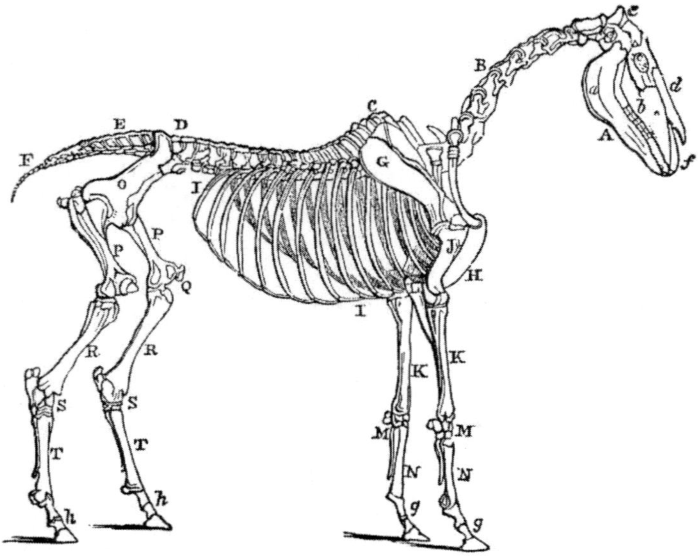

Figure 29. Skeleton of a Horse. A The Head.—a The posterior maxillary or under jaw.—b The superior maxillary or upper jaw. A little lower down than the letter is a foramen, through which pass the nerves and blood-vessels which chiefly supply the lower part of the face.—c The orbit, or cavity containing the eye.—d The nasal bones, or bones of the nose.—e The suture dividing the parietal bones below from the occipital bones above.—f The inferior maxillary bone, containing the upper incisor teeth.—B The Seven Cervical Vertebræ, or bones of the neck.—C The Eighteen Dorsal Vertebræ, or bones of the back.—D The Six Lumbar Vertebræ, or bones of the loins.—E The Five Sacral Vertebræ, or bones of the haunch.—F The Caudal Vertebræ, or bones of the tail, generally about fifteen.—G The Scapula, or shoulder-blade.—H The Sternum, or fore-part of the chest.—I The Costæ or ribs, seven or eight articulating with the sternum, and called the true ribs; and ten or eleven united together by cartilage, called the false ribs.—J The Humerus, or upper bone of the arm.—K The Radius, or upper bone of the arm.—L The Ulna, or elbow. The point of the elbow is called the Olecranon.—M The Carpus, or knee, consisting of seven bones.—N The metacarpal bones. The larger metacarpal or cannon or shank in front, and the smaller metacarpal or splint bone behind.—g The fore pastern and foot, consisting of the Os Suffraginis, or the upper and larger pastern bone, with the sesamoid bones behind, articulating with the cannon and greater pastern; the Os Coronæ, or lesser pastern; the Os Pedis, or coffin-bone; and the Os Naviculare, or navicular, or shuttle-bone, not seen, and articulating with the smaller pastern and coffin-bones.—h The corresponding bones of the hind-feet.—O The Haunch, consisting of three portions: the Ilium, the Ischium, and the Pubis.—P The Femur, or thigh.—Q The stifle joint with the Patella.—R The Tibia, or proper leg bone; behind is a small bone called the fibula—S The Tarsus, or hock, composed of six bones. The prominent part is the Os Calcis, or point of the Hock.—T The Metatarsals of the hind leg.

Glanders

This is one of the most alarming. The first and most marked symptom is a discharge from the nostrils of a peculiar character. The disease produces inflammation there and in the windpipe, and in aggravated cases passes down to the lungs, which are soon destroyed.

It is propagated by contagion, by exposure in humid stables, and is induced by hereditary indisposition and great exhaustion. Youatt says, there is not a disease which may not lay the foundation for glanders. The poison resides in the nasal discharge, not in the breath. When exposed to it, the mangers should be thoroughly scraped, washed with soap and water, and afterwards with chloride of lime. All the clothing and harness, which may have received any of the contagious matter, must be thoroughly cleansed and baked.

The best preventives are dry, clean, and well-ventilated stables, proper exercise, green food in summer, and roots in winter.

The disease may be arrested in its early stages, by turning the animal on a dry pasture, but it is liable to return on subsequent confinement. Iodine has lately been announced as a remedy, but of the certainty of its effects, we are not aware.

It is generally considered incurable, and when thoroughly seated, it may be deemed an act both of humanity and economy to terminate the existence of its victim at once. This course becomes a duty, from the fact that many grooms, by their attendance on glandered horses, have been affected, and though the disease is, in their case, more manageable, yet it is frequently fatal.

Farcy

This is intimately connected with glanders, and the diseases frequently run into each other.

Lampas

This consists in the swelling of the bars of the mouth to a level or even above the teeth. It may occur from inflammation of the gums; shedding of the teeth; a febrile tendency, and from over-feeding or want of exercise. It will generally subside by low dieting and proper exercise; or it may be at once relieved by lancing the bars with a sharp penknife.

Poll-Evil

This arises from some contusion or injury to the head, which produces a swelling that eventually suppurates. The inflammation may be abated, in its earliest stages, by a blister; and in its later, by bleeding, physic, and cold lotions applied to the part. If these are ineffectual, and the swelling continues, it should be hastened by poultices, and warm, stimulating lotions; and when fully formed, the tumor must be opened, so as to permit all the matter to run out. Repeated applications of salt will sometimes cure it.

Heaves

All those affections, distinguished in the English veterinary works as pneumonia or inflammation of the lungs, chronic cough, thick and broken wind, consumption, &c., are popularly designated as heaves. To some or all of these the horse may have an hereditary or constitutional tendency. Their incipient stages are also induced by a sudden transition from heat to cold, and sometimes from cold to close and hot stables; and by a chilly wind or damp stables, especially after severe exercise. Feeding on musty, dry hay, or on straw, will produce an irritation which may lead to heaves.

Inflammation of the lungs is frequently dangerous, and requires the immediate and full use of the lancet. After the inflammation is decidedly allayed, by copious bleeding, small doses of aloes may be given, and frequent injections of warm soap and water, which should be omitted the moment the feces become soft and approaching to the fluid state. Blistering the sides and brisket, to be often repeated if otherwise ineffectual, must be resorted to. Convalescence should be followed with sedative medicines, and during all the severe stages of the disease, withhold all food, except light gruels after protracted abstinence. As health returns, put the animal out to grass.

Inflammation of the lungs is sometimes succeeded by a chronic cough, and the other maladies enumerated. When firmly seated, it is incapable of removal. Its effects can be alleviated, and with suitable food and treatment, the horse may be made to do much moderate labor for many years, but he can never become sound or sustain great exertion. Equal and proper temperature, moist, stimulating food, and especially carrots or potatoes, and moderate exercise, (but never on a full stomach,) and dry, clean stables, are all the remedies that can be prescribed.

Catarrh or Horse Distemper

This sometimes attacks the horse in the spring or fall, and is shown by soreness and swelling in the glands of the throat, a cough; difficulty of swallowing; discharging at the nose; and general prostration. It is seldom fatal, if properly managed. Give light bran-mashes, purge thoroughly, and keep warm. If he is violently attacked, he may be bled while fever exists; and blisters or setons may be applied, to reduce the swelling, if extreme. The disease is contagious, and the animal should be at once placed where he cannot communicate it.

Spasmodic Coli

Symptoms.—The attack of colic is usually very sudden. There is often not the slightest warning. The horse begins to shift his posture, look round at his flanks, paw violently, strike his belly with his feet, and crouch in a peculiar manner, advancing his hind limbs under him; he will then suddenly lie, or rather fall down, and balance himself upon his back, with his feet resting on his belly. The pain now seems to cease for a little while, and he gets up and shakes himself, and begins to feed; the respite, however, is but short—the spasm returns more violently—every indication of pain is increased—he heaves at the flanks, breaks out into a profuse perspiration, and throws himself more recklessly about. In the space of an hour or two, either the spasms begin to relax, and the remissions are of longer duration, or the torture is augmented at every paroxysm; the intervals of ease are fewer and less marked, and inflammation and death supervene. The pulse is but little affected at the commencement, but it soon becomes frequent and contracted, and at length is scarcely tangible.

Among the *causes* of colic are, the drinking of cold water when the horse is heated. There is not a surer origin of violent spasm than this. Hard water is very apt to produce this effect. Colic will sometimes follow the exposure of a horse to the cold air or a cold wind after strong exercise. Green food, although, generally speaking, most beneficial to the horse, yet, given in too large a quantity, or when he is hot, will frequently produce gripes. Doses of aloes, both large and small, are not unfrequent causes of colic.

In some horses there seems to be a constitutional predisposition to colic. They cannot be hardly worked, or exposed to unusual cold, without a fit of it. In many cases, when these horses have died, calculi have been found in some part of the alimentary canal. Habitual costiveness and the presence of calculi, are frequent causes of spasmodic colic. The seat of colic is occasionally the duodenum, but oftener the ileum or the jejunum; sometimes, however, both the cæcum and colon are affected. Fortunately, we are acquainted with several medicines that allay these spasms; and the disease often ceases as suddenly as it appeared.

Remedies.—Turpentine is one of the most powerful remedies, especially in union with opium, and in good warm ale. A solution of aloes will be advantageously added to the turpentine and opium. If relief is not obtained in half an hour, it will be prudent to bleed, for the continuance of violent spasm may produce inflammation. Some practitioners bleed at first, and it is far from bad practice; for although the majority of cases will yield to turpentine, opium, and aloes, an early bleeding may occasionally prevent the recurrence of inflammation, or at least mitigate it. If it is clearly a case of colic, half of the first dose may be repeated, with aloes dissolved in warm water. The stimulus produced on the inner surface of the bowels by the purgative, may counteract the irritation that caused the spasm.

The belly should be well rubbed with a brush or warm cloth, but not bruised and injured by the broom-handle rubbed over it, with all their strength, by two great fellows. The horse should be walked about or trotted moderately. The motion thus produced in the bowels, and the friction of one intestine over the other, may relax the spasm, but the hasty gallop might speedily cause inflammation to succeed to colic.

Clysters of warm water, or containing a solution of aloes, should be injected. The patent syringe will here be exceedingly useful. A clyster of tobacco-smoke may be thrown up as a last resort.

When relief has been obtained, the clothing of the horse, saturated with perspiration, should be removed, and fresh and dry clothes substituted. He should be well littered down in a warm stable or box, and have bran mashes and lukewarm water for the two or three next days. Some persons give gin, or gin and pepper, or even spirit of pimento, in cases of gripes. This course of proceeding is, however, exceedingly objectionable. It may be useful, or even sufficient, in ordinary cases of colic; but if there should be any inflammation, or tendency to inflammation, it cannot fail to be highly injurious.

Flatulent Colic

This is altogether a different disease from the former. It is not spasm of the bowels, but inflation of them, from the presence of gas emitted by undigested food. Whether collected in the stomach, or small or large intestines, all kinds of vegetable matter are liable to ferment. In consequence of this fermentation gas is evolved to a greater or less extent—perhaps to twenty or thirty times the bulk of the food. This may take place in the stomach; and if so, the life of the horse is in immediate danger, for the animal has no power to expel this dangerous flatus by eructation.

The symptoms, according to Professor Stewart, are, "The horse suddenly slackening his pace, preparing to lie down, or falling down, as if he were shot. In the stable he paws the ground with his fore-feet, lies down, rolls, starts up all at once, and throws himself down again with great violence, looking wistfully at his flanks, and making many fruitless attempts to void his urine."

The treatment is considerably different from that of spasmodic colic. The spirit of pimento would be here allowed, or the turpentine and opium drink; but if the pain, and especially the swelling, do not abate, the gas, which is the cause of it, must be got rid of, or the animal is inevitably lost. This is usually, or almost invariably, a combination of hydrogen with some other gas. It has a strong affinity for chlorine. Then if some compound of chlorine—the chloride of lime—dissolved in water, is administered in the form of a drink, the chlorine separates from the lime as soon as it comes in contact with the hydrogen, and muriatic gas is formed. This gas, having a strong affinity for water, is absorbed by any fluid that may be present, and, quitting its gaseous form, either disappears, or does not retain a thousandth part of its former bulk. All this may be very rapidly accomplished, for the fluid is quickly conveyed from the mouth to every part of the intestinal canal.

Where these two medicines are not at hand, and the danger is imminent, the trochar may be used, in order to open a way for the escape of the gas. The trochar should be small, but longer than that which is used for the cow, and the puncture should be made in the middle of the right flank, for there the large intestines are most easily reached. It is only when the practitioner despairs of otherwise saving the life of the animal, that this operation should be attempted. Much of the danger would be avoided by using a very small trochar, and by withdrawing it as soon as the gas has escaped. The wound in the intestines will then probably close, from the innate elasticity of the parts.

Inflammation of the Bowels

There are two varieties of this malady. The first is inflammation of the external coats of the intestines, accompanied by considerable fever, and usually costiveness. The second is that of the internal or mucous coat, and almost invariably connected with purging. The muscular coat is that which is oftenest affected. Inflammation of the external coats of the stomach, whether the peritoneal or muscular, or both, is a very frequent and fatal disease. It speedily runs its course, and it is of great consequence that its early symptoms should be known.

Symptoms.—If the horse has been carefully observed, restlessness and fever will have been seen to precede the attack. In many cases a direct shivering fit will occur; the mouth will be hot, and the nose red. The animal will soon express the most dreadful pain by pawing, striking at his belly, looking wildly at his flanks, groaning, and rolling. The pulse will be quickened and small; the ears and legs cold; the belly tender, and sometimes hot; the breathing quickened; the bowels costive; and the animal becoming rapidly and fearfully weak.

The causes of this disease are, first of all and most frequently, sudden exposure to cold. If a horse that has been highly fed, carefully groomed, and kept in a warm stable, is heated with exercise, and has been during some hours without food, and in this state of exhaustion is suffered to drink freely of cold water, or is drenched with rain, or have his legs and belly washed with cold water, an attack of inflammation of the bowels will often follow. An overfed horse, subjected to severe and long-continued exertion, if his lungs were previously weak, will probably be attacked by inflammation of them; but if the lungs were sound, the bowels will on the following day be the seat of disease. Stones in the intestines are an occasional cause of inflammation, and colic neglected or wrongly treated will terminate in it.

Remedies.—The treatment of inflammation of the bowels, like that of the lungs, should be prompt and energetic. The first and most powerful means of cure will be bleeding. From six to eight or ten quarts of blood, in fact as much as the horse can bear, should be abstracted as soon as possible; and the bleeding repeated to the extent of four or five quarts more, if the pain is not relieved and the pulse has not become rounder and fuller. The speedy weakness that accompanies this disease, should not deter from bleeding largely. That weakness is the consequence of violent inflammation of these parts; and if that inflammation is subdued by the loss of blood, the weakness will disappear. The bleeding should be effected on the first appearance of the disease, for there is no malady that more quickly runs its course.

A strong solution of aloes should immediately follow the bleeding, but, considering the irritable state of the intestines at this period, guarded by opium. This should be quickly followed by back-raking, and injections consisting of warm water, or very thin gruel, in which Epsom salts or aloes have been dissolved; and too much fluid can scarcely be thrown up.

The horse should likewise be encouraged to drink plentifully of warm water or thin gruel; and draughts, each containing a couple of drachms of dissolved aloes, with a little opium, should be given every six hours, until the bowels are freely opened. It will now be prudent to endeavor to excite considerable external inflammation, as near as possible to the seat of internal disease, and therefore the whole of the belly should be blistered.

In a well-marked case of this disease, no time should be lost in applying fomentations, but the blister at once resorted to. The tincture of Spanish flies, whether made with spirits of wine or turpentine, should be thoroughly rubbed in. The legs should be well bandaged, in order to restore the circulation in them, and thus lessen the flow of blood to the inflamed part; and, for the same reason, the horse should be warmly clothed, but the air of the stable or box should be cool.

No grain or hay should be allowed during the disease, but bran mashes, and green food if it can be procured. The latter will be the best, and may be given without the slightest apprehension of danger. When the horse begins to recover, a handful of grain may be given two or three times in the day; and, if the weather is warm, he may be turned into a paddock for a few hours in the middle of the day. Clysters of gruel should be continued for three or four days after the inflammation is beginning to subside, and good hand-rubbing applied to the legs.

The Second Variety of Inflammation of the Bowels

Affects the internal or mucous coat, and is generally the consequence of physic in too great quantity, or of an improper kind. The purging is more violent and continues longer than was intended; the animal shows that he is suffering great pain; he frequently looks round at his flanks; his breathing is laborious, and the pulse is quick and small, and the mouth is hot and the legs and ears are warm.

Remedies.—Unless the purging is excessive, and the pain and distress great, the surgeon should hesitate at giving any astringent medicine at first; but he should plentifully administer gruel or thin starch, or arrow-root, by the mouth and by clyster, removing all hay and grain, and particularly green food. He should thus endeavor to soothe the irritated surface of the bowels, while he permits all remains of the purgative to be carried off. If, however, twelve hours have passed, and the purging and the pain remain undiminished, he should continue the gruel, adding to it chalk, catechu, and opium, repeated every six hours. As soon as the purging begins to subside, the astringent medicine should be lessened in quantity, and gradually discontinued.

Bleeding will rarely be necessary, unless the inflammation is very great, and attended by symptoms of general fever. The horse should be warmly clothed, and placed in a comfortable stable, and his legs should be hand-rubbed and bandaged.

Violent purging, and attended with much inflammation and fever, will occur from other causes. Green food will frequently purge; and a horse worked hard upon it will sometimes scour.

The remedy is change of diet, or less labor. Young horses will often be strongly purged, without any apparent cause. Astringents should be used with much caution here. It is probably an effort of nature to get rid of something that offends. A few doses of gruel will assist in effecting this purpose, and the purging will cease without astringent medicine.

Many horses that are not well ribbed home (having too great space between the last rib and the hip-bone) are subject to purging, if more than usual exertion is required from them. They are recognised by the term of *washy* horses. They are often free and fleet, but destitute of continuance. They should have rather more than the usual allowance of grain, with beans, when at work. A cordial ball, with catechu and opium, will often be serviceable either before or after a journey.

Physicking

When a horse comes from grass to dry food, or from the cool, open air to a heated stable, a dose, or even two doses, of physic may be useful to prevent the tendency to inflammation, which is the necessary consequence of so sudden and great a change. To a horse that is becoming too fat, or has surfeit, grease, or mange, or that is out of condition from inactivity of the digestive organs, a dose of physic is often most serviceable. A horse should be carefully prepared for the action of physic. Two or three bran mashes given on that or the preceding day, are far from sufficient when a horse is about to be physicked, whether to promote his condition, or in obedience to custom. Mashes should be given until the dung becomes softened. A less quantity of physic will then suffice, and it will more quickly pass through the intestines, and be more readily diffused over them.

Five drachms of aloes, given when the dung has thus been softened, will act more effectually and much more safely than seven drachms, when the lower intestines are obstructed by hardened feces. On the day on which the physic is given, the horse should have walking exercise, or may be gently trotted for a quarter of an hour twice in the day; but after the physic begins to work, he should not be moved from his stall. Exercise would then produce gripes, irritation, and, possibly, dangerous inflammation. A little hay may be put into the rack. As much mash should be given as the horse will eat, and as much water, with the coldness of it taken off, as he will drink. If, however, he obstinately refuses to drink warm water, it is better that he should have it cold, than to continue without taking any fluid; but in such case he should not be suffered to take more than a quart at a time, with an interval of at least an hour between each draught.

When the purging has ceased, or *the physic is set*, a mash should be given once or twice every day until the next dose is taken, between which and the *setting* of the first, there should be an interval of a week. The horse should recover from the languor and debility occasioned by the first dose, before he is harassed by a second. Eight or ten tolerably copious motions, will be perfectly sufficient to answer every good purpose, although the groom may not be satisfied unless double the quantity are procured. The consequence of too strong purgation will be, that weakness will hang about the animal for several days or weeks, and inflammation will often ensue from the over-irritation of the intestinal canal.

Long-continued custom has made aloes the almost invariable purgative of the horse, and very properly so; for there is no other at once so sure and so safe. The Barbadoes aloes, although sometimes very dear, should alone be used. The dose, with a horse properly prepared, will vary from four to seven drachms. Custom has assigned the form of a ball to physic, but good sense will in due time introduce the solution of aloes, as acting more speedily, effectually, and safely.

The only other purgative on which dependence can be placed is the croton. The farina or meal of the nut is generally used; but from its acrimony it should be given in the form of ball, with linseed meal. The dose varies from a scruple to half a drachm. It acts more speedily than the aloes, and without the nausea which they produce; but it causes more watery stools, and, consequently, more debility.

Linseed oil is an uncertain but safe purgative, in doses from a pound to a pound and a half. Olive oil is more uncertain, but safe; but CASTOR OIL, that mild aperient in the human being, is both uncertain and unsafe. Epsom salts are inefficacious, except in the immense dose of a pound and a half, and then they are not always safe.

Worms

The long white worm (*lumbricus teres*) much resembles the common earth-worm, and being from six to ten inches in length, inhabits the small intestines. It is a formidable looking animal; and if there are many of them, they may consume more than can be spared of the nutritive part of the food, or the mucus of the bowels. A tight skin, and rough coat, and tucked-up belly, are sometimes connected with their presence. They are then, however, voided in large quantities.

Remedies.—A dose of physic will sometimes bring away almost incredible quantities of them. Calomel is frequently given as a vermifuge. The seldomer this drug is administered to the horse, the better. When the horse can be spared, a strong dose of physic is an excellent vermifuge, so far as the long round worm is concerned. But a better medicine, and not interfering with either the feeding or work of the horse, is emetic tartar, with ginger, made into a ball with linseed meal and treacle, and given every morning, half an hour before the horse is fed.

A smaller, darker colored worm, called the needle-worm, or *ascaris*, inhabits the larger intestines. Hundreds of them sometimes descend into the rectum, and immense quantities have been found in the cœcum. These are a more serious nuisance than the former, for they cause a very troublesome irritation about the fundament, which sometimes sadly annoys the horse. Their existence can generally be discovered by a small portion of mucus, which, hardening, is found adhering to the anus.

Remedies.—Physic will sometimes bring away great numbers of these worms; but when there is much irritation about the tail, and much of this mucus, indicating that they have descended into the rectum, an injection of linseed oil, or of aloes dissolved in warm water, will be a more effectual remedy. The tape-worm is seldom found in the horse.

Bots

While they inhabit the stomach of the horse, cannot give the animal any pain, for they have fastened on the cuticular and insensible coat. They cannot stimulate the stomach and increase its digestive power, for they are not on the digestive portion of the stomach. They cannot, by their roughness, assist the trituration or rubbing down of the food, for no such office is performed in that part of the stomach—the food is softened, not rubbed down. They cannot be injurious to the horse, for he enjoys the most perfect health when the cuticular part of his stomach is filled with them, and their presence is not even suspected until they appear at the anus.

They cannot be removed by medicine, because they are not in that part of the stomach in which medicine is usually conveyed; and if they were, their mouths are too deeply buried in the mucus for any medicine, that can be safely administered, to affect them; and, last of all, in due course of time they detach themselves, and come away. Therefore, the wise man will leave them to themselves, or content himself with picking them off when they collect under the tail and annoy the animal.

Wind-Galls

In the neighborhood of the fetlock, there are occasionally found considerable enlargements, oftener on the hind-leg than the fore one, which are denominated wind-galls. Between the tendons and other parts, and wherever the tendons are exposed to pressure or friction, and particularly about their extremities, little bags or sacs are placed, containing and suffering to ooze slowly from them, a mucous fluid to lubricate the parts. From undue pressure, and that most frequently caused by violent action and straining of the tendons, or often from some predisposition about the horse, these little sacs are injured. They take on inflammation, and sometimes become large and indurated. There are few horses perfectly free from them. When they first appear, and until the inflammation subsides, they may be accompanied by some degree of lameness; but otherwise, except when they attain a great size, they do not interfere with the action of the animal, or cause any considerable unsoundness. The farriers used to suppose that they contained wind—hence their name, wind-galls; and hence the practice of opening them, by which dreadful inflammation was often produced, and many a valuable horse destroyed.

Remedies.—A slight wind-gall will scarcely be subjected to treatment; but if these tumors are numerous and large, and seem to impede the motion of the limb, they may be attacked first by bandage. The roller should be of flannel, and soft pads should be placed on each of the enlargements, and bound down tightly upon them. The bandage should also be wetted with warm water, two or three times a day, for half an hour each time. The wind-gall will often diminish or disappear by this treatment, but will too frequently return, when the horse is again hardly worked.

A blister is a more effectual, but too often temporary remedy. Firing is still more certain, if the tumors are sufficiently large and annoying to justify our having recourse to measures so severe; for it will not only effect the immediate absorption of the fluid, and the reduction of the swelling, but, by contracting the skin, will act as a permanent bandage, and therefore prevent the reappearance of the tumor. The iodine and mercurial ointments have occasionally been used with advantage, in the proportion of three parts of the former to two of the latter.

Inflammation of the Fetlock

The fetlock-joint is a very complicated one, and from the stress which is laid on it, and its being the principal seat of motion below the knee, it is particularly subject to injury. There are not many cases of sprain of the back-sinew, that are not accompanied by inflammation of the ligaments of this joint; and numerous supposed cases of sprain higher up, are simple affections of the fetlock. It requires a great deal of care, and some experience, to distinguish the one from the other. The heat about the part, and the point at which the horse least endures the pressure of the finger, will be the principal guides.

Remedy.—Occasionally, by the application of cooling lotions, the inflammation may be subdued, but at other times, the horse suffers dreadfully, and is unable to stand. A serious affection of the fetlock-joint demands prompt treatment.

Cutting

The inside of the fetlock is often bruised by the shoe or the hoof of the opposite foot. Many expedients used to be tried to remove this; the inside heel has been raised and lowered, and the outside raised and lowered; and sometimes one operation has succeeded, and sometimes the contrary; and there was no point so involved in obscurity or so destitute of principles to guide the practitioner.

Remedy.—The most successful remedy and that which in the great majority of cases supersedes all others, is Mr. Turner's shoe, of equal thickness from heel to toe, and having but one nail, and that near the toe, on the inside of the shoe; care being taken that the shoe shall not extend beyond the edge of the crust, and that the crust shall be rasped a little at the quarters.

Sprain of the Coffin-Joint

The proof of this is when the lameness is sudden, and the heat and tenderness are principally felt round the coronet.

Remedy.—Bleeding at the toe, physic, fomentation, and blisters, are the usual means adopted. This lameness is not easily removed, even by a blister; and if removed, like sprains of the fetlock and of the back-sinews, it is apt to return, and finally produce a great deal of disorganization and mischief in the foot. Sprain of the coffin-joint sometimes becomes a very serious affair. Not being always attended by any external swelling, and being detected only by heat round the coronet, the seat of the lameness is often overlooked by the groom and the farrier; and the disease is suffered to become confirmed before its nature is discovered.

Ringbone

This is a deposite of bony matter in one of the pasterns, and usually near the joint. It rapidly spreads, and involves not only the pastern-bones, but the cartilages of the foot, and spreading around the pasterns and cartilages, thus derives its name. When the first deposite is on the lower pastern, and on both sides of it, and produced by violent inflammation of the ligaments of the joints, it is recognised by a slight enlargement, or bony tumor on each side of the foot, and just above the coronet.

Horses with short upright joints, and with small feet and high action, are oftenest, as may be supposed, the subjects of this disease, which is the consequence either of concussion or sprain of the pastern-joints. It is also more frequent in the hind foot than the fore, because, from the violent action of the hind legs in propelling the horse forward, the pasterns are more subject to ligamentary injury behind than before; yet the lameness is not so great there, because the disease is confined principally to the ligaments, and the bones have not been injured by concussion; while, from the position of the fore limbs, there will generally be in them injury of the bones to be added to that of the ligaments.

Remedy.—In its early stage, and when recognised only by a bony enlargement on both sides of the pastern-joint, or in a few cases on one side only, the lameness is not very considerable, and it is not impossible to remove the disease by active blistering, or by the application of the cautery; but there is so much wear and tear in this part of the animal, that the inflammation and the disposition to the formation of bone rapidly spread. The pasterns first become connected together by bone instead of ligament, and thence results what is called an anchylosed or fixed joint. From this joint the disease proceeds to the cartilages of the foot, and to the union between the lower pastern, and the coffin and navicular bones. The motion of these parts likewise is impeded or lost, and the whole of the foot becomes one mass of spongy bone.

Enlargement of the Hock

First, there is inflammation, or *sprain of the hock-joint generally*, arising from sudden violent concussion, by some check at speed, or over-weight, and attended with enlargement of the whole joint, and great tenderness and lameness.

Remedy.—This, however, like all other diffused inflammations, is not so untractable as an intense one of a more circumscribed nature, and by rest and fomentation, or, perchance, firing, the limb recovers its action, and the horse becomes fit for ordinary work. The swelling, however, does not always subside. Enlargement, spread over the whole of the hock-joint, remains.

A horse with an enlarged hock must always be regarded with suspicion. In truth, he is unsound. The parts, altered in structure, must be to a certain degree weakened. The animal may discharge his usual work during a long period, without return of lameness; but if one of those emergencies should occur when all his energies require to be exerted, the disorganized and weakened part will fail. He may be ridden or driven moderately for many a year without inconvenience, yet one extra hard day's work may lame him forever.

Curb

There are often injuries of particular parts of the hock-joint. *Curb* is an affection of this kind. It is an enlargement at the back of the hock, three or four inches below its point. It is either a strain of the ring-like ligament which binds the tendons in their place, or of the sheath of the tendons; oftener, however, of the ligament than of the sheath. Any sudden action of the limb of more than usual violence may produce it, and therefore horses are found to throw out curbs after a hardly-contested race, an extraordinary leap, a severe gallop over heavy ground, or a sudden check in the gallop.

Young horses are particularly liable to it, and horses that are cow-hocked, whose hocks and legs resemble those of the cow, the hocks being turned inwards, and legs forming a considerable angle outwards. This is intelligible enough; for in hocks so formed, the annular ligament must be continually on the stretch, in order to confine the tendon. Curbs are generally accompanied by considerable lameness at their first appearance, but the swelling is not always great. They are best detected by observing the leg sidewise.

Remedies.—The first object in attempting the cure, is to abate inflammation, and this will be most readily accomplished by cold evaporating lotions, frequently applied to the part. Equal portions of spirit of wine, water, and vinegar, will afford an excellent application. It will be almost impossible to keep a bandage on. If the heat and lameness are considerable, it will be prudent to give a dose of physic, and to bleed from the subcutaneous vein, whose course is near it; and whether the injury is of the annular ligament, or the sheath of the tendon, more active means will be necessary to perfect a cure.

Either a liquid blister should be rubbed on the part, consisting of a vinus or turpentine tincture of cantharides, and this daily applied until some considerable swelling takes place; or, what is the preferable plan, the hair should be cut off, and the part blistered as soon as the heat has been subdued. The blister should be repeated, until the swelling has disappeared, and the horse goes sound.

In severe cases it may be necessary to fire; but a fair trial, however, should be given to milder measures. If the iron is used, it should be applied in straight lines.

There are few lamenesses in which absolute and long-continued rest is more requisite. It leaves the parts materially weakened, and, if the horse is soon put to work again, the lameness will frequently return. No horse that has had curbs, should be put even to ordinary work in less than a month after the apparent cure; and, even then he should very gradually resume his former habits.

A horse with a curb, is manifestly unsound, or generally condemned as unsound. Curb is also a hereditary complaint; and therefore a horse that has once suffered from it, should always be regarded with suspicion, especially if either of the parents have exhibited it.

Bone Spavin

This is an affection of the bones of the hock-joint. Spavined horses are generally capable of slow work. They are equal to the greater part of the work of the farm, and therefore they should not always be rejected by the small farmer, as they may generally be procured at a small price. These horses are not only capable of agricultural work, but they generally improve under it. The lameness in some degree abates, and even the bony tumor to a certain degree lessens. There is sufficient moderate motion and friction of the limb to rouse the absorbents to action, and cause them to take up a portion of the bony matter thrown out, but not enough to renew or prolong inflammation. It cannot be said that the plow affords a *cure* for spavin, but the spavined horse often materially improves while working at it. For fast work, and for work that must be regularly performed, spavined horses are not well calculated; for this lameness behind produces great difficulty in rising, and the consciousness that he will not be able to rise without painful effort, occasionally prevents the horse from lying down at all; and the animal that cannot rest well, cannot long travel far or fast.

Remedies.—The treatment of spavin is simple enough, but far from being always effectual. The owner of the horse will neither consult his own interest, nor the dictates of humanity, if he suffers the chisel and mallet, or the gimlet, or the pointed iron, or arsenic to be used; yet measures of considerable severity must be resorted to. Repeated blisters will usually cause either the absorption of the bony deposite, or the abatement or removal of the inflammation of the ligaments, or, as a last resource, the heated iron may be applied.

Swelled Legs

The fore-legs, but oftener the hind ones, and especially in coarse horses, are sometimes subject to considerable enlargement. Occasionally, when the horse does not seem to labor under any other disease, and sometimes from an apparent shifting of disease from other parts, the hind legs suddenly swell to an enormous degree from the hock, and almost from the stifle to the fetlock, attended by a greater or less degree of heat, and tenderness of the skin, and sometimes excessive and very peculiar lameness. The pulse likewise becomes quick and hard, and the horse evidently labors under considerable fever. It is acute inflammation of the cellular substance of the legs, and that most sudden in its attack, and most violent n its degree, and therefore attended by the effusion of a considerable quantity of fluid into the cellular membrane. It occurs in young horses, and in those which are over-fed and little exercised.

Remedies.—Fomentation, diuretics, or purgatives, or, if there is much fever, a moderate bleeding, will often relieve the distention almost as suddenly as it appeared. Mild cases will generally yield to their influence; but, if the animal has been neglected, the treatment must be decisive. If the horse is in high condition, these should be preceded or accompanied by bleeding; but if there are any symptoms of debility, bleeding would only increase the want of tone in the vessels.

Horses taken from grass and brought into close stables, very speedily have swelled legs, because the difference of food and increase of nutriment rapidly increase the quantity of the circulating fluid, while the want of exercise takes away the means by which it might be got rid of. The remedy here is sufficiently plain.

Swelled legs, however, may proceed from general debility. They may be the consequence of starvation, or disease that has considerably weakened the animal; and these parts, being farthest from the centre of circulation, are the first to show the loss of power by the accumulation of fluid in them. Here the means of cure would be to increase the general strength, with which the extremities would sympathize. Mild diuretics and tonics would therefore be evidently indicated.

Horses in the spring and fall are subject to swelled legs. The powers of the constitution are principally employed in providing a new coat for the animal, and the extremities have not their share of vital influence. Mingled cordials and diuretics are indicated here; the diuretic to lessen the quantity of the circulating fluid, and the cordial to invigorate the frame.

Grease

Swelled legs, although distinct from grease, is a disease that is apt to degenerate into it. Grease is a specific inflammation of the skin of the heels, sometimes of the fore-feet, but oftener of the hinder ones. Bad stable management is the true cause of it. Grease is a local complaint. The heel should be well but gently washed with soap and water, and as much of the scurf detached as is easily removable. An ointment should be applied, to supple, cool, and heal the part.

When cracks appear, the mode of treatment will depend on their extent and depth. If the cracks are deep, with an ichorous discharge and considerable lameness, it will be necessary to poultice the heel. A poultice of linseed meal will be generally effective, unless the discharge is thin and offensive, when an ounce of finely-powdered charcoal should be mixed with the linseed meal; or a poultice of carrots, boiled soft and mashed. After the chaps or cracks have healed, the legs will sometimes continue gorged and swelled. A flannel bandage, evenly applied over the whole of the swelled part, will be very serviceable; or should the season admit of it, a run at grass, particularly spring grass, should be allowed.

The feeding should likewise vary with the case, but with these rules, which admit of no exception; that green food should be given, and more especially carrots, when they are not too expensive; and mashes, if the horse will eat them; and never the full allowance of grain.

Walking exercise should be resorted to as soon as the horse is able to bear it, and this by degrees may be increased to a gentle trot.

From bad stable-management at first, and neglect during the disease, a yet worse kind of grease occasionally appears. The ulceration extends over the skin of the heel and the fetlock, and a fungus springs from the surface of both, highly sensible, bleeding at the slightest touch, and interspersed with scabs. By degrees portions of the fungus begin to be covered with a horny substance, protruding in the form of knobs, and collected together in bunches. These are known by the name of *grapes*. A fœtid and very peculiar exudation, proceeds from nearly the whole of the unnatural substance. The horse evidently suffers much, and is gradually worn down by the discharge. The assistance of a veterinary surgeon is here indispensable.

Some horses are more subject to grease than others, particularly draught-horses, both heavy and light, but particularly the former, and if they have no degree of blood in them. It was the experience of this, which partly contributed to the gradual change of coach and other draught-horses to those of a lighter breed. In the great majority of cases, grease arises from mismanagement and neglect.

Every thing that has a tendency to excite inflammation in the skin of the heel is a cause of grease. Therefore want of exercise is a frequent source of this disease. When high feeding is added to irregular or deficient exercise, the disease is evidently still more likely to be produced. Want of cleanliness in the stable is a fruitful source of grease. When the heels are imbedded in filth, they are weakened by the constant moisture surrounding them. The absurd practice of washing the feet and legs of horses when they come from their work, and either carelessly sponging them down afterwards, or leaving them to dry as they may, is, however, the most common origin of grease.

When the horse is warmed by his work, and the heels share in the warmth, the momentary cold of washing may not be injurious, if the animal is immediately rubbed dry; yet even this would be better avoided; but to wash out the heels, and then leave them partially dry or perfectly wet, and suffering from the extreme cold that is produced by evaporation from a moist and wet surface, is the most absurd, dangerous, and injurious practice that can be imagined. It is worse, when the post-horse or the plow-horse is plunged up to his belly in the river or pond immediately after his work.

Cutting the Hair from the Heels

Custom has very properly retained the hair on our farm-horses. Nature would not have given it had it not been useful. It guards the heel from being injured by the inequalities of the plowed field, and prevents the dirt, in which the heels are constantly enveloped, from reaching, and caking on, and irritating the skin. When the horse is carefully tended after his work is over, and his legs quickly and completely dried, the less hair he has about them the better, for then both the skin and the hair can be made perfectly dry before evaporation begins, or proceeds so far as to deprive the legs of their heat. Grease is the child of negligence and mismanagement.

Setons

Are pieces of tape or cord, passed, by means of an instrument resembling a large needle, either through abscesses, or the base of ulcers with deep sinuses, or between the skin and the muscular or other substances beneath. They are retained there by the ends being tied together, or by a knot at each end. The tape is moved in the wound twice or thrice in the day, and occasionally wetted with spirits of turpentine, or some acrid fluid, in order to increase the inflammation which it produces, or the discharge which is intended to be established.

In abscesses, such as occur in the withers or the poll, and when passed from the summit to the very bottom of the swelling, setons are highly useful by discharging the purulent fluid, and suffering any fresh quantity of it that may be secreted to flow out; and, by the degree of inflammation which they excite on the interior of the tumor, stimulating it to throw out healthy granulations, which gradually occupy and fill the hollow. In deep fistulous wounds they are indispensable, for except some channel is made through which the matter may flow from the bottom of the wound, it will continue to penetrate deeper into the part, and the healing process will never be accomplished. On these accounts, a seton passing through the base of the ulcer in poll-evil and fistulous withers is so beneficial.

"Setons are sometimes useful by promoting a discharge in the neighborhood of an inflamed part, and thus diverting and carrying away a portion of the fluids which distend or overload the vessels of that part; thus, a seton is placed with considerable advantage in the cheek, when the eyes are much inflamed."—(*Youatt.*)

Founder

Or inflammation of the foot, arises from various causes; excessive exertion, great heat, and particularly when followed by drinking cold water, or overloading the stomach in any way, sudden transition from great cold to excessive heat, and change of inflammation from some other part.

Remedies.—When the attack is severe and confined to the fore-feet, Youatt recommends removing the shoe and paring the hoof as much as possible, taking four quarts of blood from each toe, placing the feet in warm water, and afterwards applying soft poultices of linseed meal to the whole foot and pastern. If this is ineffectual, take three quarts of blood from each foot the succeeding day. It may then be necessary to blister the foot and coronet. The animal should be kept on green food or light mashes, and allowed to run on grass without labor. An effectual cure has followed from taking off the shoe, and applying lard, raised to the boiling point, to every part of the foot.

Poison

From weeds, sometimes gives to horses ulcerated tongues and lips, and swollen legs and sheath. If there be much inflammation, bleeding should be resorted to, then give daily bran mashes, with Glauber salts in doses of ½ to 1½ pounds, according to the size of the horse, with half a teaspoonful of saltpetre. Washing the ulcerated parts with warm soap-suds, copperas, and sugar-of-lead may follow.

Epidemics Among Horses

Sometimes occur, producing great mortality. One of these was prevalent in the neighborhood of New York, in 1846, termed a malarious congestive fever, staggers, or apoplexy, which destroyed many valuable animals. It occurred during the heat of summer, and was principally confined to such as were at pasture. By many it was attributed to excessive heat and exposure at night. The animals that were opened, appeared sound in all respects excepting the brain, which exhibited one mass of clotted blood.

The remedy found to be the most effectual, consisted in taking about one quart of blood from the head, swathing it with cloths saturated with cold water, and giving two drachms daily of calomel. The horse should be kept in a cool stable.

A similar disease in Spain is cured by copious bleeding, and swathing the head in blankets constantly wet with hot water.

Inflammation of the Eyes

Shut up in a dark stable, and feed on fresh-cut grass and bran mashes. Bleed freely from the mouth, and give 1½ lbs. Glauber salts, 2 drachms nitre, and 15 grains tartarized antimony, dissolved in a bucket of water, which the animal will drink when thirsty. This to be repeated daily till purging is effected. If it fails, bleed from the large veins just below the eye, taking 15 to 20 ounces of blood.—(Dr. Campbell.)

The Sting of Hornets

Bees, or snakes, may be relieved by immediate external application of strong spirits of hartshorn: salt and vinegar are also good.

For Sprains

Take a mixture of one ounce sweet oil, four ounces spirits of hartshorn, half an ounce oil of thyme, and rub with it frequently. The remedies mentioned below are also effectual for sprains.

For a Bruise or Blow

Apply hot water a long time with wet cloths. Beef brine is an excellent lotion for both sprains and bruises. A veteran among horses, claims, that it will almost set a joint or heal a fracture. Wormwood or tansy lotions are also good.

Fistula

This is frequently cured by repeated applications of salt.

Wounds

Should be washed twice a day with clean, soft water, or with a little Castile soap added, and then rub with whale-oil. This answers for all seasons, keeps off flies, restores the hair, and of the original color.

Galls

Or wounds on the back from the saddle, are most effectually healed by white-lead, moistened with sweet-oil or milk. The saddle ought always to fit easily and be well padded, and it should be taken off and the animal's back washed at every baiting.

Shoeing

This is an important operation, and should never be attempted but under the supervision of an experienced person; nor ought the shoes to remain so long as to produce contraction of the hoof, which is followed by lameness and corns. They should be reset as often as every five or six weeks.

Contraction of the Foot

This is also caused by standing on the dry stable for some days. In this case, the hoof should be stopped with fresh cow-manure and clay, or with a thick felt, soaked in water, and cut to suit the foot. This is also a good application over night, for horses that have accomplished a hard day's work on a dry road.

Litter

This is not objectionable to the feet, if clean and not too damp. Some suppose this the cause of contraction, but it is the reverse. It is besides of great benefit when shook out for a bed, by inducing the horse to rest himself. He is thus enabled to do more work, and with a less expenditure of food.

Corns

In the angle between the bars and the quarters, the horn of the sole has sometimes a red appearance, and is more spongy and softer than at any other part. The horse flinches when this portion of the horn is pressed upon, and occasional or permanent lameness is produced. This disease of the foot is termed *corns*; bearing this resemblance to the corn of the human being, that it is produced by pressure, and is a cause of lameness. When corns care neglected, so much inflammation is produced in that part of the sensible sole, that suppuration follows, and to that, quittor succeeds, and the matter either undermines the horny sole, or is discharged at the coronet.

Remedies.—The cure of old corns is difficult; for as all shoeing has some tendency to produce pressure here, the habit of throwing out this diseased horn is difficult to get rid of when once contracted; recent corns, however, will yield to good shoeing. The first thing to be done is well to pare out the angle between the crust and the bars. Two objects are answered by this; the extent of the disease will be ascertained, and one cause of it removed. A very small drawing-knife must be used for this purpose. The corn must be pared out to the very bottom, taking care not to wound the sole. It may then be discovered whether there is any effusion of blood or matter underneath. If this is suspected, an opening must be made through the horn, the matter evacuated, the separated horn taken away, the course and extent of the sinuses explored, and introduce into them a saturated solution of sulphate of zinc, by means of a small syringe. Place over this dressing the common cataplasm, or the turpentine ointment, and renew the application every twenty-four hours. Three or four such applications complete a cure.

Should there be no collection of fluid, the butyr of antimony should be applied over the whole extent of the corn, after the horn has been thinned as closely as possible. The object of this is to stimulate the sole to throw out more healthy horn. In bad cases, a bar-shoe may be put on, so chambered that there shall be no pressure on the diseased part. This may be worn for one or two shoeings, but not constantly, for there are few frogs that would bear the constant pressure of the bar-shoe; and the want of pressure on the heel, generally occasioned by their use, would produce a softened and bulbous state of the heels, that would of itself be an inevitable source of lameness. Turning out to grass, after the horn is a little grown, first with a bar-shoe, and afterwards with the shoe fettered on one side, or with tips, will often be serviceable. A horse that has once had corns to any considerable extent should, at every shoeing, have the seat of corn well pared out, and the butyr of antimony applied.

An Over-Reach

Is a tread upon the heel of the coronet of the fore foot, by the shoe of the corresponding hind foot, and either inflicted by the toe, or by the inner edge of the inside of the shoe.

The preventive treatment is the bevelling, or rounding off of the inside edge or rim of the hind shoes.

The cure is the cutting away of the loose parts, the application of Friar's balsam, and protection from the dirt.

Forging, or Clicking

"Is a singular species of over-reaching. The horse, in the act of trotting, strikes the toes of the hind shoes against the fore one. This noise of the clicking is unpleasant, and the trick or habit is not altogether free from danger. It is most frequent in young horses, and is attributable to too great activity, or length of stride in the hind legs.

Remedies.—The rider may do something by keeping the head of the horse well up; but the smith may effect more by making the hind shoes of clicking horses short in the toe, and having the web broad. When they are too long, they are apt to be torn off; when too narrow, the hind foot may bruise the sole of the fore one, or may be locked fast between the branches of the fore shoe."—Youatt.

The Bearing Rein

Is a matter of much controversy; some claiming that it should be entirely abolished, while others as strenuously contend for its almost universal use. Nimrod, who is deemed perfectly competent authority, insists on its use with fast roadsters and coach-horses. With team-horses, it may generally be dispensed with, and always should be in ascending hills, as it materially diminishes their capacity for exertion. The fault in its use is its excessive tightness, and when standing, the horse ought never to be tormented with it.

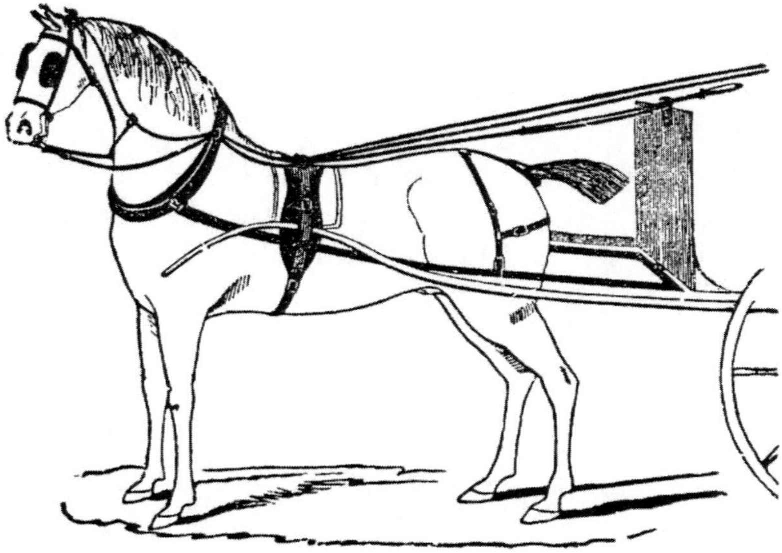

Figure 30. Safety Rein illustrated.

Directions for use of Safety Rein.—In putting on the rein for a gig, keep the buckle to the left hand, or near side; that will place the loop, which is on the middle of the rein, below the hook or head of the bridle, which prevents it from being thrown out by the motion of the horse's head. For a pair of horses, keep the two short chapes outmost, and the loops on the middle downward. For saddle, keep the buckle to the left hand.

When the rein is used either for running, rearing, kicking, or going backward, it should be applied suddenly with a strong arm, keeping up the pressure until the horse is still; it should then be relieved suddenly, at the same time motioning the horse to go on. If he is only a runaway he will obey it at once, such horses being generally of a willing, good temper. After the horse has been a few times firmly gripped with it, use it occasionally, instead of the bit-rein, to stop him on ordinary occasions; this will remind the horse of his subjection, and will accustom the rider or driver to the ready and accurate use of it in case of an emergency.

The Bit

Is a frequent cause of injury to the mouth of the horse, fretting and teasing him, and in many cases inducing permanent injury and viciousness. It should never be made annoying to the horse beyond the absolute necessity for his proper restraint.

An Unruly Stud May Be Controlled

By passing the rein from the ring on the off-side over the head and through the left ring. This gives a purchase to the groom which the horse cannot resist.

Blinds

Have for a long time been fashionable, but in few cases are necessary, while in nearly all they are decidedly injurious.

The Crib Biter

Figure 31. The Crib Biter.

This small instrument is made entirely of iron, and riveted firmly to the head-stall. It answers the threefold purpose, to prevent biting, crib-biting, and wind-sucking. All of the foregoing are bad habits for horses, for which there is no effectual cure, but in adopting the use of the above implement.

The Stable

Is an important matter connected with the proper management of horses. This should be as much as possible of a uniform temperature, cool in summer, warm in winter, and always clean, dry, and well-ventilated. But no air must be allowed to blow directly upon the animal. The horse is a native of a warm climate, and ought to be well protected against cold. The stable should be neither too light nor too dark, nor must the light ever be admitted before the eye of the horse. For judicious and extended arrangement of stables, and management of horses, the inquiring reader is referred to *Stewart's Stable Economy*.

Chapter 6

The Ass, the Mule, and the Comparative Labor of Working Animals

The Ass

Is a native of Arabia, Persia, and the central parts of Asia and Africa. Like the horse, he goes in troops and displays great natural sagacity, activity, and courage. Job says, "He scorneth the multitude of the city, neither regardeth the crying of the driver." Like the horse, too, he has from time immemorial been tamed, and become the faithful servant of man; but unlike him, he is subject to few maladies, is hardy and enduring, and subsists and even thrives on coarse and scanty forage. Thus Job says of his natural haunts, "Whose house I have made the wilderness, and the barren land his dwellings; the range of the mountains is his pasture, and he searcheth after every green thing." Xenophon, in his Anabasis, a thousand years later, says of one of the Asiatic deserts through which he passed with the army of Cyrus, "that it was full of wormwood; if any other kinds of shrubs or reeds grew there, they had all an aromatic smell; but no trees appeared. Of wild creatures, the most numerous are wild asses, which our horses sometimes chased; but the wild asses exceeded them much in speed."

Varieties

The different breeds of asses are supposed to be quite as numerous as those of the horse. Four distinct races are mentioned in the earliest scriptures. In modern times we find a similar diversity. There are two kinds in Persia, the largest a slow, heavy brute, used only for burdens; the other smaller and more spirited, and used for the saddle. In Egypt, a considerable though less marked difference exists, those near the Delta being inferior to those which are bred in Upper Egypt and Nubia. In Spain, a difference in size and spirit prevails, greater even than in Persia.

The Zebra is nearly allied in size, shape, and character to the wild ass, but his untameable ferocity has hitherto effectually bid defiance, alike to the scourges and caresses, the frowns and the favors of man.

Arabia produces some of the most spirited and hardy asses, but their size, like that of their horses, is too small for purposes of the greatest utility. The Maltese Jack is by American breeders deemed the choicest animal from which to propagate. He is evidently of Arabian descent, and possesses all the good qualities of his ancestry, with considerable additional size.

We have several varieties, all of which are imported, as there are no natives of the Western Continent. The early importations were principally made from the Azores, and Cape de Verd Islands, and were mostly of an inferior character. A superior Maltese Jack was presented to Gen. Washington, in 1787, by La Fayette, and is believed to be the first ever sent to this country. Mr. Custis describes him as of moderate size, clean-limbed, possessing great activity, the fire and ferocity of a tiger, of a dark brown and nearly black, with white belly and muzzle, and manageable only by one groom, nor then safely. He lived to a great age. His mules were all active, spirited, and serviceable, and when from stout mares, attained considerable size.

A Spanish Jack and Jennet were also presented to Washington about the same time, by the King of Spain. The first is characterized by the same authority, as a huge, ill-shapen animal, nearly 16 hands high, very large head, clumsy limbs, and to all appearance little calculated for active service; he was of a gray color, and not much valued for his mules, which were unwieldy and dull. From the Maltese Jack and Spanish Jennet, which approach the size of the large Spanish Jack, was bred a valuable animal, Compound, which partook of all the good qualities of the sire, with the weight of the dam. From him descended many of the best mules of Mount Vernon.

Many other valuable importations followed these animals, and it is believed we have for many years had as fine specimens of the ass as the world affords.

Jennets, or she-asses, are used among us principally for breeding Jacks, and of course are not numerous. They are sometimes, though seldom, bred to the horse. It is difficult to induce the horse to notice them, and the produce, which is called a hinny, is less hardy and useful than the mule. The milk of the she-ass is lighter and more digestible than that of any other animal, and in former times was in great request for invalids.

The ass is occasionally used in the cart, or as a beast of burden. Such as are employed for these purposes are generally of an inferior kind, and are only used for the lightest work. They may sometimes be seen among the fishmongers and small vegetable dealers about our city markets, but little larger than a Newfoundland dog or Shetland pony, trundling along a light cart with a wheelbarrow load. In ancient times they have been, and in foreign countries—even at the present time, they are extensively used. But the most enlightened of the moderns have adopted the mule as the proper and almost exclusive substitute for the ass; and it would show a still greater intelligence and economy, if it much more extensively took the place of the horse.

THE MULE

Is the hybrid produced by the ass with the mare. How early this animal was bred, is uncertain, but we know he was in high repute in the reign of David, near 3,000 years ago, for he was rode by Absalom, the favorite prince of Israel, on the field of battle. They have from time immemorial been bred in various parts of the East, on the borders of the Mediterranean, and throughout Spain, Portugal, and other countries, many of them being of splendid appearance and of fine qualities. In these countries, they are frequently used by the grandees and nobles, and indeed by royalty itself; and however much they may be undervalued elsewhere, when they are finely bred and trained, and richly caparisoned, they exhibit a stateliness and bearing, that few of the highest bred horses can match.

Breeding Mules in the United States

Was commenced with much spirit in some of the New England states, soon after the American revolution. The object was not to breed them for their own use, but only as an article of commerce. They were at first shipped exclusively to the West Indies; and afterwards to the South and West, for employment in the various work of the plantation. Indifferent animals, both as sires and dams, were used at first, as anything which bore the name of mule, then commanded a ready sale. The progeny were necessarily inferior brutes, and viewed with almost universal derision; and being considered the type of their race, a prejudice was excited against them, which more than half a century has not been sufficient to dispel.

Among a few thinking men at the North, they have been adopted and made highly useful in the various duties of the farm. They have been largely introduced at the South and West, but principally in the slave states, where the management of the team devolves upon the ignorant and heedless. It is there, and in other and hotter climates, that the superior merits of the mule over the horse as a laboring animal, are peculiarly manifest. In many instances they are indifferently fed, hardly worked, and greatly neglected by their drivers; yet they sustain themselves for years, in defiance of usage that would annihilate two generations of horses. Their powers have been largely increased and their merits improved, by the introduction of some of the best Maltese and Spanish Jacks, and the use of large, blood mares. The propriety of this course is seen in the value of the product; for while some of the inferior are unsaleable at $50, others of the same age, and reared under the same circumstances of keep and condition, could not be purchased for $150.

The Breeding, Rearing, and Management of Mules

Is similar to that of colts. They will be found, equally with horses, to repay generous keep and attention, by their increased and rapid growth. But they should not be pampered by high feed, as it not only has a tendency to produce disease, but to form habits of fastidiousness, which materially lessens their economical feeding in after life. The diseases to which mules are subjected, (which are always few, and if properly managed will seldom or ever occur,) require a treatment like that of horses.

The breeding from mules has sometimes been questioned, but it has been demonstrated in several instances. Neither the sexual development nor propensities are wanting, but they are seldom indulged with effect. Mr. Kilby, of Virginia, states in the Farmer's Register, that a mare mule brought two colts from a young horse, which they closely resembled. The first was a male, and died, apparently with staggers, which no treatment could arrest, at six months old. The second was a female, 16 months younger than the first, marked like the sire, being jet-black, excepting a white foot and star in the forehead, and died at a year old, after two days' illness, notwithstanding the utmost care was bestowed upon it. Successful propagation of this hybrid, however, beyond the first cross, seems to be incompatible with the fixed laws of nature.

With a view of encouraging the substitution of mules for a part of the horses now employed in American husbandry, we give the following testimony from experienced individuals, of great intelligence and careful observation.

ADVANTAGES OF MULE OVER HORSE LABOR

The official report of an agricultural committee in South Carolina, in 1824, says:—

> "The annual expense of keeping a horse is equal to his value. A horse at four years old would not often bring more than his cost. Two mules can be raised at less expense than one horse. The mule is fit for service earlier, and if of sufficient size, will perform as much labor as the horse; and if attended to when first put to work, his gait and habits may be formed to suit the owner."

Mr. Pomeroy, who used them near Boston for 30 years, and to such an extent as to have had more labor performed by them probably than any person in New England, says:—"I am convinced the small breed of mules will consume less in proportion to the labor they are capable of performing than the larger race, but I shall confine myself to the latter in my comparison, such as stand 14½ to 16 hands, and are capable of performing any work a horse is usually put to. From repeated experiments, I have found that three mules of this description, which were constantly at work, consumed about the same quantity of hay, and only one-fourth the provender, which was given to two middling-size coach-horses, only moderately worked. I am satisfied a large-sized mule will not consume more than three-fifths to two-thirds the food to keep him in good order, that will be necessary for a horse performing the same labor. The expense of shoeing a mule the year round, does not exceed one-third that of the horse, his hoofs being harder, more horny, and so slow in their growth, that shoes require no removal, and hold on till worn out; and the wear from the lightness of the animal is much less.

Mules have been lost by feeding on cut straw and corn meal; in no other instance have I known disease in them, except by inflammation of the intestines, caused by the grossest exposure to cold and wet, and excessive drinking cold water, after severe labor, and while in a high state of perspiration. It is not improbable a farmer may work the same team of mules for twenty years, without having a farrier's bill presented to him.

In my experience of thirty years, I have never found but one mule inclined to be vicious, and he might have been easily subdued while young. I have always found them truer pullers and quicker travellers, with a load, than horses. Their vision and hearing are much more accurate. I have used them in my family carriage, in a gig, and under the saddle; and have never known one to start or run from any object or noise, a fault in the horse, that continually causes the maiming and death of numerous human beings.

The mule is more steady in his draught, and less likely to waste his strength than the horse, hence more suitable to work with oxen; and as he walks faster, he will habituate them to a faster gait. In plowing among crops, his feet being small and following each other so much more in a line, he seldom treads down the ridges or crops. The facility of instructing him to obey implicitly the voice of the driver is astonishing. The best plowed tillage land I ever saw, I have had performed by two mules tandem, without lines or driver. The mule is capable of enduring labor in a temperature of heat that would be destructive to a horse.

Although a large mule will consume something over one-half the food of a horse, yet the saving in shoeing, farrying, and insurance against diseases and accidents, will amount to at least one-half. In addition, the owner may rely with tolerable certainty on the continuance of his mule capital for thirty years; whereas the horse owner must, at the end of fifteen years, look to his crops, his acres, or a bank for the renewal of his. The longevity of a mule is so proverbial, that a purchaser seldom inquires his age. Pliny mentions one 80 years old; and Dr. Rees, two in England, that reached the age of 70. I saw one performing his labor in a cane-mill in the West Indies, which the owner assured me was 40 years old. I have now a mare-mule 25 years old, that I have had in constant work for 21 years. She has often within a year taken a ton weight in a wagon to Boston, five miles, and manifests no diminution of her powers. A neighbor has one 28 years old, which he would not exchange for any horse in the country. One in Maryland, 35 years old, is now as capable of labor as at any former period."

Mr. Hood of Maryland, in the American Farmer, estimates the annual expense of a horse for 12 months, at $44, and that of a mule at $22, just half price, and his working age at more than twice that of the horse, and that too after 30 years' experience in keeping both.

A correspondent of the Baltimore Patriot, asserts that "Col. John E. Howard had a pair of mules that worked 30 years, after which they were sold to a carter in the city, and performed hard service for several years longer. Many mules 25 years old, and now in this country, perform well. Many have been at hard work for 12 or 15 years, and would now sell for $100 each. They are not subject to the colt's ailments, the glanders, heaves, yellow-water, and colic, like horses; and seldom are afflicted with spavin, ringbones, or bots; and they will not founder."

General Shelby says, "he has known mules to travel 12 miles within the hour in light harness, and has himself driven a pair 45 miles in six hours, stopping an hour by the way." Four match mules have been sold in this country for $1,000. They were of course superior animals, and made elegant coach-horses. These animals were driven 80 miles in a day without injury; and they proved a first-rate team for many years.

Mr. Ellicott, of the Patuxent Furnaces, asserts that, "out of about 100 mules at the works, we have not lost on an average one in two years. Bleeding at the mouth will cure them of nearly every disease, and by being turned out on pasture, they will recover from almost every accident. I do not recollect we have ever had a wind-broken one. They are scarcely ever defective in the hoof, and though kept shod, it is not as important as with the horse. Their skin is tougher than that of the horse, consequently they are not as much worried by flies, nor do they suffer so much with the heat of summer."

To the foregoing testimony may be added that of the late Judge Hinckley of Massachusetts; a shrewd and close observer through a long life of 84 years. He bred mules at an early day, and always kept a team of them for his farm work, much preferring them to horses for this purpose, after an experience of 50 years. He had a pair nearly 30 years old, which, with light pasturage in summer, and with a moderate supply of hay with little grain in winter, and no grooming, performed all the drudgery, though he kept his stable full of horses besides. They outlived successive generations of horses, and though the latter were often sick and out of condition, the mules never were. One from his stock, 45 years old, was sold for the same price paid for a lot of young mules, being at that mature age perfectly able to perform his full share of labor.

For the caravans that pass over the almost inaccessible ranges which form the continuation of the Rocky Mountains, and the extensive arid plains that lie between and west of them, on the route from Santa Fé to California, mules are the only beasts of burden used in these exhausting and perilous adventures. Their value may be estimated from the comparative prices of mules and horses; for while a good horse may be bought for $10 to $20, a good mule is worth $50 to $75.

Dr. Lyman, who recently passed through those regions, informs us that their caravan left Santa Fé with about 150 mules, 15 or 20 horses, all beasts of burden, and two choice blood-horses, which were led and treated with peculiar care. On the route, all the working-horses died from exhaustion and suffering; the two bloods that had been so carefully attended, but just survived; yet of the whole number of mules but 8 or 10 gave out. A mule 36 years of age was as strong, enduring, and performed as hard labor, as anyone in the caravan. When thirst compelled them to resort for successive days to the saline waters, which are the only ones furnished by those sterile plains, the horses were at once severely, and not unfrequently fatally affected; while the mules, though suffering greatly from the change, yet seldom were so much injured as to require any remission of their labor.

The mules sent to the Mexican possessions from our western states, Missouri, Tennessee, and Kentucky, are considered of much more value than such as are bred from the native (usually wild) mares. The difference probably arises, in part, from the Mexicans using jacks so inferior to most of the stock animals used by the citizens of those states.

Mare mules are estimated in those regions at one-third more than horse mules. The reason assigned for this is, that after a day's journey of excessive fatigue, there is a larger quantity of blood secreted in the bladder, which the female, owing to her larger passage, voids at once, and without much apparent suffering, while the male does not get rid of it, frequently, till after an hour of considerable pain. The effect of this difference is seen in the loss of flesh and strength in the male to an extent far beyond that of the female.

The method of reducing refractory mules in the northern Mexican possessions, is for the person to grasp them firmly by the ears, while another whips them severely on the fore-legs and belly.

Estimated annual saving to the United States from the employment of mules in the place of horses.—To sum up the advantages of working mules over horses, we shall have as advantages: 1. They are more easily, surely, and cheaply raised. 2. They are maintained, after commencing work, for much less than the cost of keeping horses. 3. They are not subject to many of the diseases of the horse, and to others only in a mitigated degree, and even these are easily cured in the mule. 4. They attain a greater age, and their average working years are probably twice that of the horse.

In 1840, there were reported to be 4,335,669 horses and mules in the Union, no discrimination having been made between them. Suppose the total number at the present time is 4,650,000, and that of these 650,000 are mules. If we deduct one-fourth, supposed to be required for the purposes of breed, fancy-horses, &c., we shall have 3,000,000 horses, whose places may be equally well supplied by the same number of mules. We have seen that Mr. Hood, of Maryland, estimates the expense of a working horse at $44 per annum, (not an over estimate for the Atlantic states,) while that of the mule is $22. The difference is $22, which it is proper to reduce to meet the much lower rate of keeping at the West. If we put the difference at $10, we shall find the saving in the keep, shoeing, farriery, &c., by substituting mules for the 3,000,000 horses that can be dispensed with, will be $30,000,000 per annum. But this is not all.

The working age of the horse will not exceed an average of eight years, while that of the mule is probably over sixteen. To the difference of keep, then, must be added the annual waste of the capital invested in the animal. A mule is more cheaply raised to working age than a horse, but allowing them to cost equally, we shall have the horse exhausting one-eighth of his capital annually for his decay, when the mule is using up but one-sixteenth; and if we allow $48 as the first cost of both animals, we shall find the horse wasting $6 annually for this item, while the mule deteriorates but $3, making an additional item of $9,000,000. This will give an aggregate of $39,000,000, as the annual saving to the United States by substituting good mules for three-fourths of the horses now used in this country. When will our farmers have the good sense to make this change? It may be fairly answered, when they shall prefer utility, interest, and a just taste, to a diseased fancy; for though we admit the superiority in appearance of the race of horses over mules, we deny that a bad horse looks better or even as well as a good mule; and with the same keep and attention, a good mule will outwork and outlook most horses of any breed.

The Comparative Economy of Horse and Ox Labor

This is a question which has been often discussed, and when with candor, the conclusion has generally been in favor of ox-labor. The different employments, the variety of situation, the season, and the kind of stock reared on the farm, are all questions which should be fully considered in arriving at their true comparative advantages. Most farmers would find it for their interest to keep teams of each, where there is employment for more than one; or if this be not the case, the preference should be given to that which is best suited in all respects to their particular position.

If work upon the road is required, a horse team will generally be best. Their superiority will consist principally in their greater speed; for even with a heavy load, they will be able to trot occasionally, and when driven without it, they may increase their pace to nearly double the natural gait of the ox. This will amount to a large annual saving in the time of the driver when steadily employed. The same is true when removing manures or crops on the farm to remote distances, over a smooth surface, which admits of trotting with the empty wagon. Harrowing ought always to be done with a quick team, as a violent stroke of the teeth breaks the clods and pulverizes the earth much better than when slowly dragged.

But we should assume in this comparison, that oxen shall not only be well adapted to their work by their natural formation, like the Hereford, the Devon, and others equally good, but also that they be well trained, well managed, accustomed to quick movements, and as well fed and looked after as horses. We shall then find their walk equal to a quick horse team, and that in this case the horse will have no advantage over the ox in harrowing. For plowing, the teams are on a par, as a good ox team will do as much in a day in cool weather as horses.

The situation of the farm may materially affect this estimate. In a warm climate, horses, and more especially mules, would be more serviceable than oxen, as they are capable of enduring much greater heat with impunity. If the farm be small and convenient to market, the labor may, in general, be best accomplished by oxen, as little travelling will be required. So, too, if the land be stony or rough, the plowing and harrowing will be more kindly and patiently done by oxen than by spirited horses. Other considerations will suggest themselves as affecting the comparative economy of this labor.

The time of work is to be fully considered. If much and heavy work be required in summer, as is often the case in plowing extensive wheat farms, horses are to be preferred; yet if the ox-team be started at early dawn, and worked briskly four or five hours, and then turned out to rest with a supply of suitable food, they may again commence when the extreme heat has abated, and accomplish a day's work that few horses will exceed. During the season of muddy roads, the horse, with his broad, compact foot, and longer leg, has a decided advantage over the ox. If the ox draws by the yoke, (which on the whole is the best mode,) he is liable to a sore neck when working in wet or snowy weather, and at such times he is overmatched by his competitor.

The kind of stock raised on the farm has an important bearing on this question. Some farms are devoted to rearing horses, and some exclusively to rearing cattle. These occasionally remain on hand after they are fit for market, from the want of a profitable demand. They can then be employed not only without injury, but in consequence of the thorough training thus secured, with positive benefit to their future value.

Even if intended for the shambles, the well-developed ox may advantageously be put to light work at three, after which it may be gradually increased till he is six or eight, and during all this time he will be improving. After doing an early spring's work, he may then be turned on to good pasture, and if followed with proper stall-feeding, he will in the latter part of the winter or spring yield a tender, better-flavored, and more profitable carcass, than can be procured by any other mode of fattening.

The first cost of oxen is less than that of horses, and they are at all times cheaply reared on the coarser herbage of the farm. The expense of working-gear, tackle, and shoeing, is much less than with horses. They are subject to fewer diseases, and these are more within the reach of ordinary medicines. The cost of food is also less, and while the horse is depreciating, the ox is increasing in value till eight or nine years old.

Accidents are less frequent with oxen, from their slower movements; and when they occur, the ox may be turned out to fatten, and still be worth as much for this purpose as for the yoke. A permanent injury to the horse is perhaps a total loss of the beast, with a large farrier's bill in addition, for which there is nothing to liquidate it but the hide.

The small farmer can make out a most serviceable team, by putting a single horse before a yoke of cattle. If well trained, they will soon accommodate themselves to each other's pace, and work as advantageously together as an entire team of either class would do alone.

Bulls are frequently put to the draught, and when they have not other services that fully test their powers, they cannot be better employed. Heifers and cows are sometimes worked, but hitherto they have not been used to any extent in this country. In the absence of other animals, they might perform light work to advantage, but severe labor would stint their growth or impair their milk beyond the benefit derived from it. The spayed heifer is an exception to the foregoing remark, and by many is esteemed even more useful than an ox of equal weight.

We have no definite statements of the comparative money value of the labor of oxen and horses. But in England repeated trials have been made, and while some have discovered no advantage in the employment of oxen over horses, others have proved them decidedly superior. One Anglesey farmer found in an experience of three years, with 12 horses and 20 oxen, which accomplished an equal amount of work, that he had saved by the latter, $1150.

The foregoing facts prove the subject to be one of sufficient importance, to justify the closest investigation of every farmer to determine for himself the comparative value of ox, horse, or mule labor.

Chapter 7

SWINE

The hog is a cosmopolite of almost every zone, though his natural haunts, like those of the hippopotamus, the elephant, the rhinoceros, and most of the thick-skinned animals, are in warm climates. They are most abundant in China, the East Indies, and the immense range of islands which extends over the whole Southern and Pacific Oceans; but they are also numerous throughout Europe, from its southern coast to the Russian dominions within the Arctic.

Figure 32. The Wild Boar.

In the United States, swine have been an object of attention since its earliest settlement, and whenever a profitable market could be found for pork abroad, it has been exported to the full extent of the demand. For near twenty years following the commencement of the general European wars, soon after the organization of our national government, it was a comparatively large article of commerce; but from that time, exports have not been justified to any extent, till within the last two years, since which, a material reduction in the British import duty on pork, lard, and hams, has again brought it up as a prominent article of trade with that country. The recent use which has been made of the carcass in converting it into lard oil, has still further increased its consumption.

Swine are reared in very part of the Union, and when properly managed, always at a fair profit. At the extreme North; in the neighborhood of large markets; and on such of the Southern plantations as are particularly suited to sugar or rice, they should not be raised beyond the number required for the consumption of the coarse or refuse food produced. Swine are advantageously kept in connection with a dairy or orchard, as with little additional food besides what is thus afforded, they can be put into good condition for the butcher. But it is on the rich bottoms and other lands of the West, where Indian corn is raised in profusion, and at small expense, that they can be reared in the greatest numbers and yield the largest profit. The Sciota, Miami, Wabash, Illinois, and other valleys, and extensive tracts in Kentucky, Tennessee, Missouri, and some adjoining states, have for many years taken the lead in the production of swine; and it is probable the climate and soil, which are peculiarly suited to their rapid growth, as well as that of their appropriate food, will enable them forever to remain the leading pork-producers of the North American continent.

BREEDS OF SWINE

The breeds cultivated in this country are numerous, and like our native cattle, they embrace many of the best, and a few of the worst to be found among the species. Great attention has for many years been paid to their improvement in the Eastern states, and nowhere are there better specimens than in many of their yards. This spirit has rapidly extended West and South; and among most of the intelligent farmers who make them a leading object of attention, on their rich corn grounds, swine have attained a high degree of excellence. This does not consist in the introduction and perpetuity of any distinct races, so much as in the breeding up to a desirable size and aptitude for fattening, from such meritorious individuals of any breed, or their crosses, as come within their reach.

Figure 33, represents an English breed of hogs, a century or more ago: though coarse and slouch-eared, it is yet the portrait of a tolerable hog, and far before many of the swine that still maintain their ascendency in various parts of the European continent. This breed is nearly extinct, having been crossed successively by the Chinese and other good breeds, thus diminishing the size and materially improving its thrift and tendency to fattening. We have few such animals in the United States, though we have many that are worse.

Figure 33. Old English Hog.

The Byefield, some 30 years ago, was a valuable hog in the Eastern states, and did much good among the species generally. They are white, with fine curly hair, well made and compact, moderate in size and length, with broad backs, and at 15 months attaining some 300 to 350 lbs. net.

The Bedford or Woburn is a breed originating with the Duke of Bedford, on his estate at Woburn, and brought to their perfection, probably, by judicious crosses of the China hog, on some of the best English swine. A pair was sent by the duke to this country, as a present to Gen. Washington, but they were dishonestly sold by the messenger in Maryland, in which state and Pennsylvania they were productive of much good at an early day, by their extensive distribution through different states. Several other importations of this breed have been made at various times, and especially by the spirited masters of the Liverpool packet ships, in the neighborhood of New York. They are a large, spotted animal, well made, and inclining to early maturity and fattening. They are an exceedingly valuable hog, but are nearly extinct both in England and this country, as a breed.

The Leicesters are a large, white hog, generally coarse in the bone and hair, great eaters, and slow in maturing. Some varieties of this breed differ essentially in these particulars, and mature early on a moderate amount of food. The crosses with smaller compact breeds, are generally thrifty, desirable animals. Other large breeds deserving commendation in this country, are the large Miami white, the Yorkshire white, and the Kenilworth, each frequently attaining, when dressed, a weight of 600 to 800 lbs.

The Chinese is among the smaller varieties, and without doubt is the parent stock of the best European and American swine. They necessarily vary in appearance, size, shape, and color, from the diversity in the style of breeding, and the various regions from which they are derived.

The Figure represents the pure China pig, and is a striking likeness of many of the imported and their immediate descendants that we have seen in this country. They are too small an animal for general use, and require to be mixed with larger breeds to produce the most profitable carcass for the market. For the purpose of refining the coarse breeds, no animal has ever been so successful as this. They are fine-boned, short, and very compact, with bellies almost touching the ground, light head and ears, fine muzzle, of great docility and quietness, small feeders, and producing much meat for the quantity of food consumed.

From the rapidity with which generations of this animal are multiplied, the variety of other breeds on which they are crossed, and the treatment to which they are subjected, it is not surprising that their descendants should rapidly assume distinct features. They furnish not only a strong dash of blood in the best class of large breeds, but in such of the smaller as have any pretensions to merit, they constitute the greater part of the improvement. Such are the Neapolitan, the Essex half-black, the Grass breed, and some others.

Figure 34. China Hog.

Figure 35. Berkshire Hog.

The Berkshires are an ancient English breed, formerly of large size, slow feeders, and late in maturing. Their color was a buff or sandy ground, with large black spots, and the feet, lower part of the legs, and tuft on the tail, buff. The latter color has given place, in most of the modern age, to white in the same parts. This variation, with the more important ones of early maturity and good feeding properties, are by Professor Low ascribed to a Chinese cross, which has added the only characteristic in which they were before deficient.

They were first introduced and reared as a distinct breed in this country by Mr. Brentnall, of Orange Co., and Mr. Hawes, of Albany, N. York. In their hands, and those of other skilful breeders, their merits were widely promulgated. No other breeds have been so extensively diffused in the United States, within comparatively so brief a period, as the Berkshires, since 1832, and they have produced a marked improvement in many of our former races.

They weigh variously, from 250 to 400 lbs. net, at 16 months, according to their food and style of breeding; and some full-grown have dressed to more than 800 lbs. They particularly excel in their hams, which are round, full, and heavy, and contain a large proportion of lean, tender, and juicy meat, of the best flavor.

None of our improved breeds afford long, coarse hair or bristles; and it is a gratifying evidence of our decided improvement in this department of domestic animals, that our brush-makers are under the necessity of importing most of what they use from Russia and northern Europe. This improvement is manifest not only in the hair, but in the skin, which is soft and mellow to the touch; in the finer bones, shorter head, upright ears, dishing face, delicate muzzle, and mild eye; and in the short legs, low flanks, deep and wide chest, broad back, and early maturity.

BREEDING

Swine should not be allowed to breed before 12 or 15 months old, unless the animals are large and coarse, when they may be put to it somewhat younger. Not only choice individuals, but such as are well descended, should be selected for the purpose of breeding. The sow should be in good condition, but not fat, nor approaching to it; and a proper degree of exercise is essential to the development of the fœtus and the health of the parent; for which reason, she should have an extended range connected with her pen.

The sow goes with young about 114 days. A week before the time comes round, a comfortable, quiet place should be prepared for her under cover, and well-protected from cold, if the weather be severe; or if warm, a range in a pasture with an open shed to retire to, is sufficient. Too much litter for bedding must be avoided, and no change or disturbance of the sow permitted, till two or three weeks after pigging, as the restlessness thereby produced may result in the loss of the pigs. The sow should be fed only with a small quantity of the lightest food or thin gruel, for two or three days, nor put on full feed for a week. If inclined to eat her pigs, she should be fed two or three times with raw pork or fresh meat. The pigs may be taught to crack oats or soaked corn after three weeks, and if provided with a trough inaccessible to the dam, they will soon learn to feed on milk and other food, preparatory to weaning. This may take place when they are 8 or 10 weeks old; and to prevent injury to the sow, let one or two remain with her a few days longer, and when finally removed, if her bag appears to be full, they may be allowed to drain the milk after 20 or 30 hours. The sow should be restricted to a light, dry diet for a few days at this period.

Management and Fattening

There are but two objects in keeping swine, for breeding and for slaughter, and their management is consequently simple. Those designed for breeding, should be kept in growing condition, on light food, and have every advantage for exercise. Such as are destined exclusively for fattening, ought to be steadily kept to the object.

It is the usual, though a bad practice in this country, to let spring pigs run at large for the first 15 months, with such food as is convenient; and if fed at all, it is only to keep them in moderate growth till the second autumn. They are then put up to fatten, and in the course of 60 or 90 days are fed off and slaughtered. During this brief period, they gain from 50 to 100 per cent more of dressed weight than in the 15 or 18 months preceding: nor even then do they yield a greater average weight than is often attained by choice, thrifty pigs, which have been well-fed from weaning to the age of 8 or 9 months.

Three pigs of the Bedford breed, when precisely 7½ months old, dressed 230, 235, and 238½ lbs. Two of the Berkshire and Leicester breeds, at 9 months, dressed 304 and 310 lbs. Three others of the Berkshire and Grass breeds, 7 months and 27 days old weighed 240, 250, and 257 lbs. net. Innumerable instances could be adduced of similar weights, gained within the same time, with a good breed of animals under judicious treatment. We have no one accurate account of the food consumed, so as to determine the relative profit of short or long feeding. But that an animal must consume much more in 18 or 20 months to produce the same quantity of dressed meat, which is made by others of 8 or 9 months, does not admit of a doubt.

We have seen that an ox requires but little more than double the quantity of food to fatten, that is necessary for supporting existence. If we apply this principle to swine, and state the quantity of food which will fatten the pig rapidly, to be three times as great as for the support of life, we shall find that the pig will fatten in 7 months, on the same food he would consume to keep him alive for 21. This is based on the supposition that both animals are of equal size. But the pig that matures and is slaughtered at 7 months, has only a moderate capacity for eating. During the early stages of his growth, his size and the consequent incapacity of the digestive organs, prevent the consumption of the same quantity which the larger animal requires; and his accumulating fat, his limited respiration, consequent upon the compression of his lungs, and his indisposition to exercise, all conspire to keep the consumption of food within the smallest possible limit. This result, in the absence of any experiment, must be conjectural entirely; but we believe that experiments will show, that of two thrifty pigs from the same litter, one of which is properly fed to his utmost capacity for 7 months, and the other fed with precisely double the quantity of similar food for 21 months, the first will yield more carcass and of a better and more profitable quality than the latter, which has consumed 100 per cent the most.

The food is only one item in this calculation. The oldest requires the most attention, is liable to more accidents and disease, besides the loss of interest. We are necessarily forced to the conclusion, that by far the cheapest mode of wintering pigs is in the pork-barrel. We can readily anticipate one objection to this practice, which is the want of food at the requisite season of the year to fatten them. This can be obviated, by reserving enough of the previous year's grain, to keep the animal in a rapidly thriving state, till the next crop matures sufficiently to feed.

In the rich corn regions, on its beginning to ripen, as it does in August, the fields are fenced off into suitable lots, and large herds are successively turned into them, to consume the grain at their leisure. They waste nothing except the stalks, which in that region of plenty are considered of little value, and they are still useful as manure for succeeding crops; and whatever grain is left by them, leaner droves which follow, will readily glean. Peas, early buckwheat, and apples, may be fed on the ground in the same way.

There is an improvement in the character of the grain from a few months' keeping, which is fully equivalent to the interest of the money and cost of storage. If fattened early in the season, they will consume less food to make an equal amount of flesh than in colder weather; they will require less attention; and generally, early pork will command the highest price in market.

It is most economical, to provide the swine with a fine clover pasture to run in during the spring and summer; and they ought also to have access to the orchard, to pick up all the unripe and superfluous fruit that falls. They should also have the wash of the house and the dairy, to which add meal, and sour in large tubs or barrels. Not less than one-third, and perhaps more, of the whole grain fed to swine, is saved by grinding and cooking or souring. Yet care must be observed that the souring be not carried so far as to injure the food by putrefaction. A mixture of meal and water, with the addition of yeast or such remains of a former fermentation as adhere to the side or bottom of the vessel, and exposure to a temperature between 68° and 77° will produce immediate fermentation.

In this process there are five stages. The *saccharine*, by which the starch and gum of the vegetables, in their natural condition, are converted into sugar; the vinous, which changes the sugar into alcohol; the mucilaginous, sometimes taking the place of the vinous, and occurring when the sugar solution, or fermenting principle is weak, producing a slimy, glutinous product; the acetic, forming vinegar, from the vinous or alcoholic stage; and the putrefactive, which destroys all the nutritive principles and converts them into a poison. The precise point in fermentation when the food becomes most profitable for feeding, has not yet been satisfactorily determined; but that it should stop short of the putrefactive, and probably the full maturity of the acetic, is certain.

The roots for fattening animals ought to be washed, and steamed or boiled; and when not intended to be fermented, the meal may be scalded with the roots. A small quantity of salt should be added. Potatoes are the best roots for swine; then parsnips; orange or red carrots, white or Belgian; sugar beets; mangel-wurzel; ruta-bagas; and the white turnips, in the order mentioned. The nutritive properties of turnips are diffused through so large a bulk, that we doubt if they can ever be fed to fattening swine with advantage; and they will barely sustain lift when fed to them uncooked.

There is a great loss in feeding roots to fattening swine, without cooking. When unprepared grain is fed, it should be on a full stomach, to prevent imperfect mastication, and consequent loss of the food. It is better indeed to have it always before them. The animal machine is an expensive one to keep in motion, and it should be the object of the farmer to put his food in the most available condition for its immediate conversion into fat and muscle.

Swine ought to be kept perfectly dry and clean, and provided with a warm shelter, to which they can retire at pleasure. This will greatly hasten the fattening and economize the food. They thrive better and are generally less subject to disease, when long confined in yards, by having a clear running stream always accessible, to wallow in. This is one of the best preventives of vermin and cutaneous diseases. A hog ought to have three apartments, one each for sleeping, eating, and evacuations, of which the last may occupy the lowest, and the first the highest level, so that nothing shall be drained, and as little carried into the first two as possible. They must be regularly fed three times a day, and if there is a surplus, it should be removed at once. If they are closely confined in pens, give them as much charcoal twice a week as they will eat. This corrects any tendency to disorders of the stomach. Rotten wood is an imperfect substitute for charcoal.

Graves, scraps, or cracklings, as they are variously called, the residuum of rough lard or tallow, after expressing the fat, are a good change and an economical food. Some animal food, although not essential, is always acceptable to swine. When about to finish them off, many feed for a few weeks on hard corn. This is proper when slops or indifferent food has been given, and meal cannot be conveniently procured; but when fattened on sound roots and meal, it is a wasteful practice, as the animal thus falls behind his accustomed growth. It is better to give him an occasional feed of the raw grain, for a change, and to sharpen his appetite.

The products furnished by the carcass of swine are numerous. Every part of the animal is used for food, and it admits of a far greater variety of preparation for the table, than any other flesh. From the remotest antiquity to the present time, and in every grade of barbarous and civilized life, it has been esteemed as one of the choicest delicacies of the epicure.

Lard-oil (oleine) has, within a few years, given to pork a new and profitable use, by which the value of the carcass is greatly increased. At some of the large pork-packing depots of the West, one-third of the whole quantity has been thus disposed of. This has withdrawn a large amount of pork from the market, and prevented the depression which must otherwise have occurred.

Where the oil is required, the whole carcass, after taking out the hams and shoulders, is placed in a tub having two bottoms, the upper one perforated with holes, on which the pork is laid, and then tightly covered. Steam, at a high temperature, is then admitted into the tub, and in a short time all the fat is extracted and falls upon the lower bottom. The remaining mass is bones and scraps. The last is fed to pigs, poultry, or dogs, or affords the best kind of manure. The bones are either used for manure, or are converted into animal charcoal, worth about three cents per pound, which is valuable for various purposes in the arts. When the object is to obtain lard of a fine quality, the animal is first skinned, and the adhering fat carefully scraped off. The oily, viscid matter of the skin is thus avoided. When tanned, the skin makes a valuable leather. An aggregate weight of 1790 lbs. from four well-fattened animals, after taking out the hams and shoulders, say about 400 lbs., gave within a fraction of 1200 lbs. of the best lard.

Stearine and Oleine.—Lard and all fatty matters consist of three principles, of which stearine contains the stearic and margaric acids, both of which, when separated, are solid, and used as inferior substitutes for wax or spermaceti candles. The other, oleine, is fluid at a low temperature, and in American commerce, is known as lard-oil. It is very pure, and extensively used for machinery, lamps, and most of the purposes for which olive or spermaceti oils are used.

CURING HAMS AND PORK

After dressing, the carcass should be allowed to hang till perfectly drained and cool, when it may be cut up and salted. The usual way is to pack the pork in clean salt, adding brine to the barrel when filled. But it may be dry salted, by rubbing it in thoroughly on every side of each piece, with a strong leather rubber firmly secured to the palm of the right hand. The pieces are then thrown into heaps and sprinkled with salt, and occasionally turned till cured; or it may at once be packed in dry casks, which are occasionally rolled to bring the salt into contact with every part.

HAMS AND SHOULDERS

May be cured in the same manner, either dry or in pickle, but with differently arranged materials. The following is a good pickle for 200 lbs. Take 14 lbs. of Turk Island salt; ½ lb. of saltpetre; 2 qts. of molasses, or 4 lbs. of brown sugar, with water enough to dissolve them. Bring the liquor to the scalding point, and skim off all the impurities which rise to the top. When cold, pour it upon the ham, which should be perfectly cool but not frozen, and closely packed; and if not sufficient to cover it, add enough pure water for this purpose. Some extensive packers in Cincinnati and elsewhere, who send choice hams to market, add pepper, allspice, cinnamon, nutmegs, or mace and cloves.

The hams may remain six or eight weeks in this pickle, then hung up in the smoke-house, with the small end down, and smoked from 10 to 20 days, according to the quantity of smoke. The fire should not be near enough to heat the hams. In Holland and Westphalia, the fire is made in the cellar, and the smoke carried by a flue into a cool, dry chamber. This is undoubtedly the best method of smoking. The hams should at all times be dry and cool, or their flavor will suffer. Green sugar-maple chips are best for smoke; next to them are hickory, sweet-birch, corn-cobs, white-ash, or beech.

The smoke-house is the best place to keep hams till wanted. If removed, they should be kept cool, dry, and free from flies. A canvass-cover for each, saturated with lime, which may be put on with a whitewash brush, is a perfect protection against flies. When not to be kept long, they may be packed in dry salt, or even in sweet brine, without injury. A common method is to pack in dry oats, baked sawdust, &c.

Diseases of Swine

Mortifying as the fact may be to human pride, it is nevertheless certain, that the internal arrangements, the viscera, digestive organs, omnivorous propensities, and the general physiological structure of the hog and the bear, more nearly resemble man, than any other animal. Many of their diseases may therefore be expected to be a modification of those of the human species, and require a similar treatment.

Pulmonary Affections, Colds, Coughs, and Measles

To each of these, swine are peculiarly liable, and, as with most other evils, prevention of disease in swine is more easy and economical than cure. A dry warm bed, free from winds or storms, and suitable food, will most effectually prevent any injuries, or fatal attacks. The hog has little external covering to protect him against cold. Nature has provided this immediately within the skin, in the deep layer of fat which surrounds the full, plump hog. Fat is one of the best non-conductors of heat, and the pig which is well-fed bids defiance to the intense cold, which would produce great suffering, and consequent disease, in the ill-conditioned animal. By the observance of a proper medium between too much fat or lean, for the store or breeding swine, and providing them with comfortable beds and proper feed, nearly all diseases will be avoided.

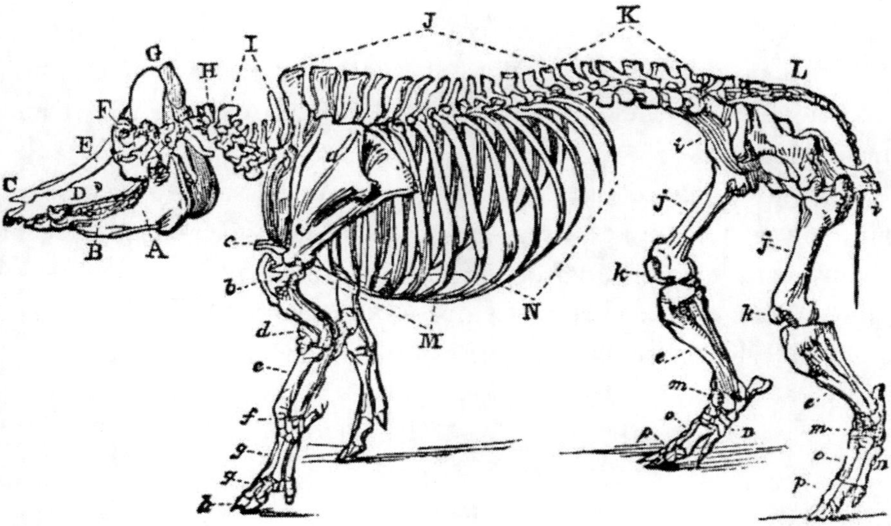

Figure 36. Skeleton of a Pig. A Maxilla inferior, vel posterior; lower jaw.—B Dentes; the teeth.—C Ossa nasi; the nasal bones.—D Maxilla superior, vel anterior; upper jaw.—E Os frontis; the frontal bone.—F Orbiculus; the orbit or socket of the eye.—G Os occipitis; the occipital bone.—H Atlas; the first vertebra of the neck.—I Vertebræ colli, vel cervicales; the vertebræ of the neck.—J Vertebræ dorsi, vel dorsales; the vertebræ of the back.—K Vertebræ lumborum, vel lumbales; the vertebræ of the loins.—L Ossa coccygis; the bones of the tail.—a Scapula; the shoulder-blade.—b Humerus; the round shoulder-bone.—c Sternum; the breastbone.—d Ulna; the elbow.—e Radius; the bone of the fore-arm.—f Os naviculare: the navicular bone.—g Phalanges vel ossa pedis; the first and second bones of the foot.—h Phalanges, vel ossa pedis; the bones of the hoof.—i Pelvis, (ossa innorninata;) the haunch bones.—j Os femoris; the thigh-bone.—k Patella; the stifle-bone.—l Tibia; the upper bone of the leg—m Tarsus, (one of which is the (N) os calcis;) the hock-bones.—n Os naviculare; the navicular bone.—o Digiti, vel phalanges, (ossa pedis;) the first digits of the foot.—p Digiti, vel phalanges, (ossa pedis;) the second digits of the foot.

For Coughs and Inflammation of the Lungs, bleeding should immediately be resorted to, after which give gentle purges of castor oil, or

Epsom salts; and this should be followed with a dose of antimonial powders—2 grains, mixed with half a drachm of nitre.

For Costiveness or loss of appetite, sulphur is an excellent remedy, given in a light mess.

Itch may be cured by anointing with equal parts of lard and brimstone. Rubbing-posts, and a running stream to wallow in are preventives.

The Kidney Worm is frequently fatal; and always produces weakness of the loins and hind legs, usually followed by entire prostration. A pig thus far gone, is hardly worth the trouble of recovering, even where practicable.

Preventives, are general thrift, a range in a good pasture, and a dose of half a pint of wood-ashes every week or fortnight in their food. A small quantity of saltpetre, spirits of turpentine, or tar, will effect the same object. When attacked, apply spirits of turpentine to the loins, and administer calomel carefully; or give half a tablespoonful of copperas daily for one or two weeks.

Blind Staggers

This is generally confined to pigs, and manifests itself in foaming at the mouth, rearing on their hind legs, champing and grinding their teeth, and apparent blindness. The proper remedies are bleeding and purging freely, and these frequently fail. Many nostrums have been suggested, but few are of any utility. It is important to keep the issues on the inside of the fore-legs, just below the knee, thoroughly cleansed.

The tails of young pigs frequently drop or rot off, which is attended with no further disadvantage to the animal than the loss of the member. The remedies are, to give a little brimstone or sulphur in the food of the dam; or rub oil or grease daily on the affected parts. It may be detected by a roughness or scabbiness at the point where separation is likely to occur.

Bleeding.—The most convenient mode, is from an artery just above the knee, on the inside of the fore-arm. It may be drawn more copiously from the roof of the mouth. The flow of blood may usually be stopped, by applying a sponge or cloth with cold water.

The diseases of swine, though not numerous, are formidable, and many of them soon become fatal. They have not been the subject of particular scientific study, and most of the remedies applied, are rather the result of casual or hap-hazard suggestion, than of well-digested inference, from long-continued and accurate observation.

Chapter 8

FARM-DOGS

No grazing farm is complete without one or more intelligent, well-trained dogs, adapted to the various wants of their owners; and the general taste has made their presence almost universal in every rural household. The dog is peculiarly the friend of man. Many other animals have a temporary though feeble attachment to him, which seems the result rather of constant companionship, or the selfishness of dependence, than any well-settled affection towards a master.

The dog alone, of all the brute creation, seems capable of a disinterested, self-sacrificing affection; and this, united with his usefulness and adaptedness to all climes and countries, has made him a favorite in every quarter of the globe. Since this animal is the habitual tenant of the farm, and, when suited to his peculiar duties, can be made of great utility by the assistance he is capable of affording in its management, we deem it entirely appropriate to our subject to indicate such of the species as are deserving the farmer's attention. Discarding all ideas of fancy or sportsmanship, and looking to utility alone, we may safely affirm that the farmer needs only such as may be found in the four breeds of the Newfoundland, the Shepherd's and Drover's dog, and the Terrier.

Figure 37. Newfoundland Dog.

THE NEWFOUNDLAND DOG

This dog, of which we give a portrait, is always above medium height, and frequently is of the largest size. He is long-haired and shaggy, and has a thick coat of fine, soft fur, beneath the outer covering, which is almost impenetrable by water. His color is most frequently black; often spotted and partially flecked or grayish; and occasionally buff. The Newfoundland is of the Spaniel family, but derives its name from the island where it has been bred for centuries, to the great advantage of its inhabitants. There are two varieties: the large, used in the north, called the Labrador; and the smaller, more docile and intelligent, of the south, called the St. John's. They are employed by the islanders, and the people of the neighboring coast, in drawing their sleds and carts loaded with fish, wood, &c. They aid them in various ways in their fishing operations; they are strong, courageous, and watchful; and with slight training, they are scarcely inferior to the best hunting-dogs in pursuing the wild game that abounds in those high northern latitudes. These estimable qualities, coupled with their uniform good-nature, have always made them favorites with the farmer.

The Newfoundland is an excellent watch-dog; sagacious in discriminating between a friend and a foe, and with courage and strength to follow out his prompt and judicious conclusions. He is easily trained for the drover, to whom he is frequently a great assistant; and with a scent sufficiently acute to pursue game, he is readily broken in as a useful companion to the sportsman. He can also be made serviceable in the various duties of the farm: destroying noxious vermin, taking the cattle and horses to the field or water, drawing a light load, churning the butter, &c. It is true, he has not all the sagacity of the Poodle, whose intelligence approaches nearer to human reason than any other of the brute creation. But if he has not that quick apprehension, which too often leads, as in the case of forward children, to the attainment of every worthless accomplishment and the indulgence of every loaferish habit, he seems to have a sedate, well-formed judgment, which makes all his wit available for some useful purpose. He is unsurpassed as a water-dog; and his courageous efforts, wherever an opportunity has been afforded, in rescuing numberless human beings from a watery grave, together with his unswerving fidelity and devotion, commend him as the prince of the canine family.

THE SHEPHERD'S DOG

This animal, of which we give a beautiful portrait on the next page, of the long-haired Scottish breed, belongs to the same family as the Newfoundland and Poodle, which embraces the most intelligent and useful of the canine species. There are two classes of these dogs, which differ widely in their size and characteristics.

The larger is of great size and courage, and when protected by a stout leather collar studded with spikes, is a full match for the wolf. These are used by the Spanish and Mexican shepherds, on their wild sierras, as effective guards against the attacks of all marauders, and are essentially the same race as the far-famed dogs of Mount St. Bernard. They are not sufficiently gentle for guides, and the shepherds who employ them rely on some well-trained wethers or goats to lead the flock at their call. Some have been imported into this country, but on account of their headstrong and ferocious character, and occasional depredations upon the flocks, they have been found unsuited to our wants, except on the borders of the wilderness.

The Colley or Scottish sheep-dog, the English, and those extensively used upon the continent, differ much in their form and appearance, but agree in their intelligence, docility, and usefulness. They are of medium size, with a sharp nose, broad forehead, and small upright ear; they are both shaggy and smooth-haired, with a bushy tail, and much hair about the neck; variously colored, though more frequently inclined to black or darkly spotted and gray; and one branch of the family is entirely destitute of a tail. They possess an instinctive sagacity for the management of sheep; and in company with a well-trained dog, under the direction of the shepherd, they soon become entirely competent to the control of the flock. They perceive his wishes by a word or sign, and with almost the speed of the greyhound, dart off to execute them. Accounts of their performances have been frequently related, which seem almost incredible to those unacquainted with their peculiar character. The following anecdote, often told by the gifted poet, Mr. James Hogg, more generally known by the soubriquet of the Ettrick Shepherd, will show their capacity more fully than any description.

"On one night, a large flock of lambs that were under the Ettrick Shepherd's care, frightened by something, scampered away in three different directions across the hills, in spite of all that he could do to keep them together. 'Sirrah,' said the shepherd, 'they're a' awa!' It was too dark for the dog and his master to see each other at any considerable distance, but Sirrah understood him, and set off after the fugitives. The night passed on, and Hogg and his assistant traversed every neighboring hill in anxious but fruitless search for the lambs; but he could hear nothing of them nor of the dog, and he was returning to his master with the doleful intelligence that he had lost all his lambs. 'On our way home, however,' says he, 'we discovered a lot of lambs at the bottom of a deep ravine called the Flesh Cleuch, and the indefatigable Sirrah standing in front of them, looking round for some relief, but still true to his charge. We concluded that it was one of the divisions which Sirrah had been unable to manage, until he came to that commanding situation. But what was our astonishment when we discovered that not one lamb of the flock was missing! How he had got all the divisions collected in the dark, is beyond my comprehension. The charge was left entirely to himself from midnight until the rising sun; and, if all the shepherds in the forest had been there to have assisted him, they could not have effected it with greater promptitude. All that I can say is, that I never felt so grateful to any creature under the sun as I did to my honest Sirrah that morning.'"

Figure 38. Shepherd's Dog.

They are quiet and good-natured, never inclined to roam or neglect their duties, and as little disposed to injure the animals intrusted to their keeping. They have almost the intelligence of the shepherd in discerning the vagaries of the flock, and ten times his efficiency in driving it. No extensive sheep-walks, unless closely hemmed in by impassable fences, should be without one or more of these useful animals.

The Drover's Dog

This animal is shown in the annexed figure. He is closely allied to the sheep-dog, from which he derives all his intelligence and capacity, differing only in being somewhat larger and heavier, which is essential to his controlling the sturdier bullocks under his charge. His additional size is acquired by crossing with some of the stouter races, such as the Newfoundland or the pointer, and even the bull-dog and large shaggy terrier have sometimes been resorted to for a strain of that indomitable courage and game, which is frequently requisite to the proper discharge of his duties. He requires more training than the sheep-dog, as his peculiar instincts are rather to the management of the flock than the herd; but when fairly broken in, he is equally expert in its management. The drover's dog may also be useful for watching, if crossed with a reference to this object, which the sheep-dog seldom is.

The Terrier

This, in addition to the foregoing, is the only dog necessary to the farm. He is needed principally for his great sagacity and indefatigable perseverance in exterminating rats and other vermin, that frequently congregate in swarms around the farmer's premises, producing such an aggregate of annoyance and devastation. Other dogs may occasionally be good ratters, but the terrier takes to them from instinct, as the

Newfoundland does to the water, or the sheep-dog to his flock. He has great ingenuity and activity in ferreting out and capturing his prey, and whenever a fair opportunity is afforded, he seldom fails of success. The famous English terrier, Billy, on two occasions, killed 100 rats in a ring at each time, in an average of less than six and a half minutes.

The terrier is usually below the medium size, but sometimes fully comes up to or even exceeds it. He is smooth-haired or rough according to the breed, of which there are several, each claiming to be equally pure. Besides his capacity for the destruction of small game, his innate love for the sport renders him a valuable assistant in keeping off vagrant cattle from the premises; and his quick ear, habitual watchfulness, and prompt courage, fully qualify him, to the extent of his size, for an excellent watch-dog.

Figure 39. Drover's Dog.

The fancy of country residents may incline them to keep a variety of other dogs than are herein enumerated, some of which, with good training, can be made partial assistants to their masters. But it is unnecessary to specify the various breeds that may possibly be of some use on the farm, as the slight crossing they will be likely to have, equally with their opportunities and the attention bestowed upon them, will serve materially to develop or obscure their peculiar instincts. The Spaniel family and its crosses will be found to combine the greatest intelligence, fidelity, and aptitude to learn; the hound has the keenest scent and greatest endurance in the pursuit of game; while the bull-dog has the most courage, sullen ferocity, and strength. Each may occasionally be wanted for a strain of blood for some particular objects; and this is especially necessary from the bull-dog in the management of refractory cattle, or to give the terrier greater stoutness and courage.

The absurd custom of keeping from one to a dozen dogs, untrained for any valuable purpose, or supernumeraries even if capable of rendering occasional service, ought to be abandoned by every rational man. Besides the great annual cost, the danger of communicating rabies or madness is sufficient to justify a legal restraint on their numbers. The sad havoc they commit on the flocks, demands the extermination by law of every dog that is guilty, whether

> Mongrel, puppy, whelp, or hound,
> Or cur of low degree.

And even if it includes the *fides Achates*, or parlor companions,

> Tray, Blanche, and Sweetheart,

the work of extirpation should proceed, to the extent of curtailing this branch of farm-stock to its wholesome and legitimate proportions.

Chapter 9

POULTRY

Choice varieties of fowls add a pleasant feature to the farm premises. They engage the attention and sympathy of the juvenile farmers, and the time bestowed in the poultry yard keeps them from mischief: it is an agreeable and salutary relief from toil and study, and elicits the taste, the judgment, and the kindlier feelings of humanity, which are to be matured in the future accomplished breeder. When properly managed, poultry are a source of considerable profit, yielding more for the food they consume, than any other stock, although their value is not often considered. The agricultural statistics of the United States, for 1839, give us over $12,000,000 in poultry, and it probably exceeds $15,000,000 at the present time. It is estimated by McQueen, that the poultry of England exceeds $40,000,000, and yet McCulloch says she imports 60,000,000 eggs annually from France, (McQueen states it at near 70,000,000;) and from other parts of the continent, 25,000,000; besides 80,000,000 imported from Ireland. Poultry, then, ceases to be an unimportant object of agricultural attention, and assumes its appropriate place among the other staples of the farmer.

HENS

Are the most numerous and profitable, and the most generally useful of the feathered tribe. The hen is peculiarly an egg-producing bird. She has the same predisposition for laying, that the cow has for secreting milk. Some breeds are better adapted for this object than others: but in all that have ever come within our notice, the proper food and circumstances are alone wanting, to produce a reasonable quantity of eggs.

The *egg* consists of three distinct parts; the shell, the white, and the yolk. A good-sized egg will weigh 1,000 grains, of which about 107 are shell, 604 are white, and 289 are yolk. Of the shell, 97 per cent is carbonate of lime, 1 per cent phosphate of lime and magnesia, and 2 per cent albumen. The white consists of 12 per cent of albumen, 2.7 of mucus, 0.3 of salts, and 85 of water. The yolk has about 17.4 per cent of albumen, 28.6 of yellow oil, 54 of water, with a trace of sulphur and phosphorus.

The foregoing are the constituents of eggs, which have been formed when the bird has free access to the various articles which constitute her natural food. But they vary with circumstances. When full-fed and denied all access to lime, she will form an egg without the shell, and deliver it enclosed in the membrane or sack which always surrounds the white, when covered by the shell. When scantily fed, they will frequently lay; but from a deficiency of nutriment, the egg will be meager and watery, and possess but a small portion of the nutritious qualities peculiar to them.

To produce the largest number of good eggs, several conditions are important; and they must especially have an abundance of the right kind of food. This is the most readily obtained in part from animal food. In warm weather, when they have a free range, they can generally supply their wants in the abundance of insects, earth-worms, and other animal matters within their reach. The large proportion of albumen contained in their eggs, requires that much of their food should be highly nitrogenized, and when they cannot procure this in animal matter, it must be given in grains containing it.

If to the usual qualities of hens, a breed of peculiar elegance, of graceful form, and beautiful plumage, be added, together with entire adaptation to the economical purposes required, good layers and good carcass, we have a combination of utility, luxury, and taste in this bird, which should commend them as general favorites. They can everywhere be kept with advantage, except in dense cities. A hen that costs a shilling or two, if provided with a suitable range, will consume 30 or 40 cents worth of food, and produce from 80 to 150 eggs per annum, worth three or four times the cost of feed and attention.

The Food

Of hens may consist of different kinds of grain, either broken, ground, or cooked; roots, and especially boiled potatoes, are nutritious and economical; green herbage as clover and most of the grasses, chickweed, lettuce, cabbage, &c., will supply them with much of their food, if fresh and tender.

Figure 40, is a Food fountain. The grain is placed in the hopper, which is closely covered, and the grain falls into the bottom below. It is accessible on four sides by spring doors, which are thrown open by the weight of the fowl on the connecting spring. One is shown as opened by the fowl in stepping up to feed. This is a protection against dirt and vermin.

Though not absolutely essential to them, yet nothing contributes so much to their laying, as unsalted, animal food. This is a natural aliment, as is shown by the avidity with which they pounce on every fly, insect, or earth-worm which comes within their reach. It would not of course pay to supply them with valuable flesh, but the blood and offal of the slaughter-houses, refuse meat of all kinds, and especially the scraps or cracklings to be had at the inciters' shops, after soaking for a few hours in warm water, is one of the best and most economical kinds of food. Such with boiled meal is a very fattening food. Grain is at all times best for them when cooked, as they will lay more, fatten quicker, and eat much less when fed to them in this state; and it may be thus used unground, with the same advantage to the fowls as if first crushed, as their digestive organs are certain to extract the whole nutriment. All grain is good for them, including millet, rice, the oleaginous seeds, as the sun-flower, flax, hemp, &c. It is always better to afford them a variety of grains where they can procure them at their option, and select as their appetite craves.

They are also fond of milk, and especially when it has become curdled; and indeed scarcely any edible escapes their notice. They carefully pick up most of the waste garbage around the premises, and glean much of their subsistence from what would otherwise become offensive; and by their destruction of innumerable insects and worms, they render great assistance to the gardener. Of course their ever-busy propensity for scratching, is indiscriminately indulged just after the seeds have been sown and while the plants are young, which renders it necessary that they be confined in some close yard for a time; yet this should be as capacious as possible.

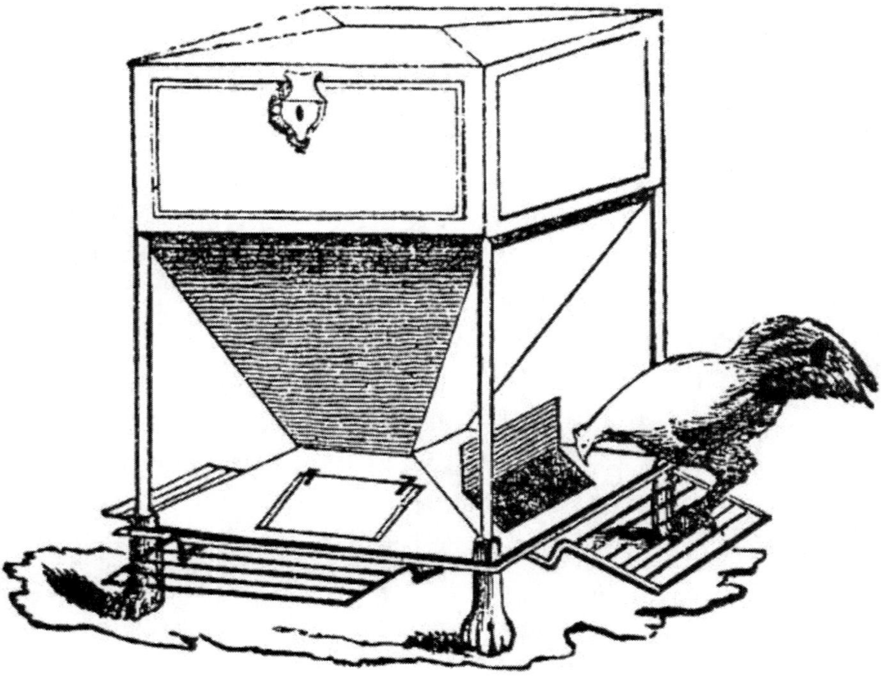

Figure 40. Food Fountain.

Figure 41. Water Fountain.

Water is placed in the cask as represented in the Figure, and it is then closely stopped, except an opening through a tube leading into a vessel below. As the water is exhausted from this, it descends from the cask above, and a supply is thus at all times within reach of the poultry.

Their food is better when given to them warm, not hot, and there should always be a supply before them to prevent gorging. It is better to be placed on shelves or suspended boxes or hoppers, which are variously and cheaply constructed, to keep it clean and out of the reach of rats. Besides their food, hens ought to be at all times abundantly supplied with clean water, egg or pounded oyster shells, old mortar or slacked lime. If not allowed to run at large, where they can help themselves, they must also be furnished with gravel to assist their digestion; and a box or bed of ashes, sand, and dust, is equally essential to roll in for the purpose of ridding themselves of vermin.

Figure 42. Poultry House.

The Hen-House

May be constructed in various ways to suit the wishes of the owner, and when tastefully built it is an ornament to the premises, It should be perfectly dry throughout, properly lighted, and capable of being made tight and warm in winter, yet afford all the ventilation desirable at any season. In this, arrange the nests in boxes on the sides, in such a manner as to humor the instinct of the hen for concealment when she resorts to them. When desirable to set the hen, these nests may be so placed as to shut out the others, yet open into another yard or beyond the enclosure, so that they can take an occasional stroll and help themselves to food, &c. This prevents other hens laying in their nests, while setting; and it may be easily managed, by having their boxes placed on the wall of the building, with a moveable door made to open on either side at pleasure. Hens will lay equally well without a nest-egg, but when broken up, they ramble off and form new nests, if they are not confined. They will lay if kept from the cock, but it is doubtful if they will thus yield as many eggs. Hens disposed to set at improper times, should be dismissed from the common yard, so as to be out of reach of the nests, and plentifully fed till weaned from this inclination.

Figure 43 represents an egg-hatcher or Eccalobeon, made of different sizes, with shelves so arranged as to hold from 200 to 800 eggs without touching each other. The outer box is a non-conductor, so as to retain the heat conveyed to every part by water tubes, connected by a reservoir below, the bottom of which is heated by the flame from a spirit-lamp. The temperature is indicated by a thermometer on the door inside, which should be made equal to that of the hen, say about 106° Fahrenheit. Her natural temperature is somewhat elevated by the feverish condition of the bird at the period of incubation.

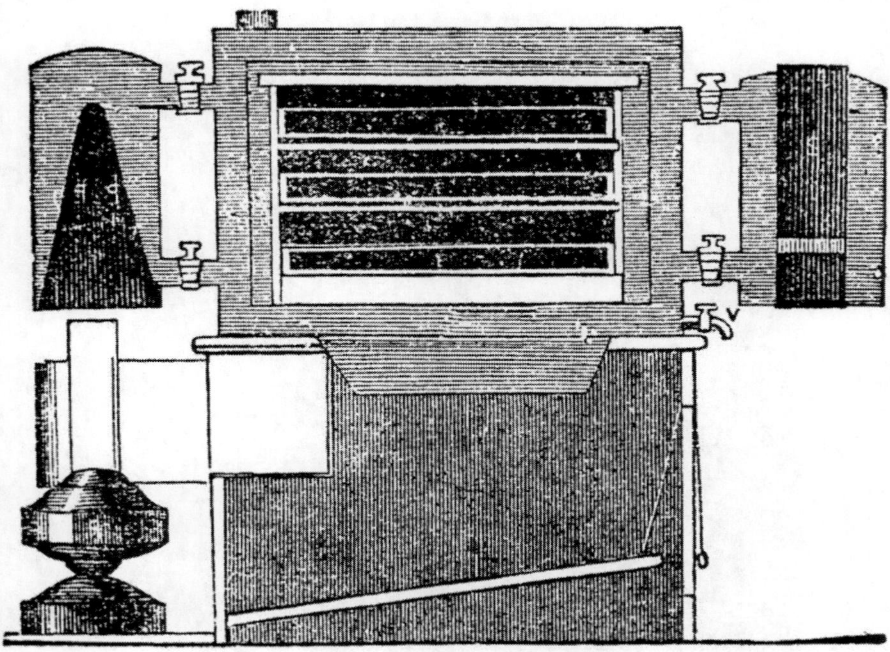

Figure 43. Egg-Hatcher, or Eccalobeon.

Chickens require to be kept warm and dry, for a few days after hatching, and they may be fed with hard-boiled eggs, crumbs of bread or pudding, and milk or water, and allowed to scratch in the gravel in front of the hen, which should be confined in a coop for the first three or four weeks. After this, they may be turned loose, when they will thrive on anything the older ones eat. Many use them for the table when they are but a few weeks old; but they are unfit for this purpose, till they have attained full maturity.

The white-legs are preferred by some, from the whiteness and apparent delicacy of the meat; but the yellow-legged are the richest and most highly-flavored. The color of the feathers does not seem to affect the quality of the flesh or their character for laying. If we consider the chemical principles of the absorption and retention of heat, we should assume the white coat to be best, as it is coolest in summer when exposed to the sun, and warmest in winter. Yet some of the white breeds are delicate, and do not bear rough usage or exposure.

Figure 44. The Dorking.

VARIETIES

These differ materially in their sizes, shapes, and colors.

The Dorking is esteemed one of the best, being large, well-formed and hardy, good layers and nurses, and yielding an excellent carcass. They are both white and speckled, and generally have five toes.

The Poland is both white and black, with a large tuft, generally of white feathers, on the head. They are of good size, and excellent layers, but are seldom inclined to set, which makes them peculiarly desirable for such as wish eggs only.

The Dominique is a speckled fowl, of barely medium size, compact, hardy, good layers, and valuable for the table. The Bucks county fowls, heretofore principally reared near Philadelphia, possess but moderate pretensions to notice, except in their immense size, a brace of capons having been fattened to 19¼ lbs. when dressed.

Figure 45. The Bantam.

The Bantam is but little larger than a pigeon, and is usually of a pure white, but is sometimes speckled. It is generally feathered to the toes, but may be bred with clean legs. It is very domestic, and a pleasant little bird around the premises, and is not unprofitable. The Game cock is of medium weight, and yields good flesh, but is a poor layer, and an undesirable tenant for the farm-yard. Besides these, there are many fanciful varieties, as the Creeper, with excessively short legs; the Rumpless, without a tail; the Frizzled, with irregular feathers turned towards the head; the Silky or Merino fowl, with brown or buff down, instead of feathers; the Negro, with its black crest, wattles, skin, legs, and feathers; the Java and Cochin China, of great size; several varieties of the Top-knot, and others.

The Diseases of Hens

These are not numerous or complicated, and may be mostly avoided by proper treatment and food, which are indicated with sufficient minuteness in the foregoing observations.

Gapes or Pip is generally owing to drinking unwholesome or dirty water. Remove the white blister on the tip of the tongue, and wash with sharp vinegar, diluted with warm water; or compel the bird to swallow a large lump of fresh butter, mixed with Scotch snuff. It has been cured by opening the mouth and forcing a pigeon feather, with a tuft of the feathers left on the end, (the others having been stripped off,) down the windpipe, and gently turning it as withdrawn, to be repeated the following day if necessary. This detaches large numbers of a slender red worm, collected in the larynx of the throat, which impedes respiration and swallowing. A little spirits of turpentine mixed with the food is a preventive; as are also clean, whitewashed premises, and good food. After these attacks, feed for a few days with light food, soaked bran and cabbage, or lettuce chopped fine.

Roup, Catarrh, or swelled head, is shown by feverish symptoms, swollen eyelids, frequently terminating in blindness, rattling in the throat, and temporary strangulation. These are accompanied with a highly offensive watery discharge, from the mouth and nostrils, loss of appetite, and much thirst. They should be placed near the fire; their head bathed in warm Castile soap-suds, or milk and water. Stimulating food, as flour or barley-meal, mustard and grated ginger, mixed and forced down the throat, Boswell says, has been effectual in their speedy restoration. This, like many other diseases, is contagious, and when it appears, the bird should be at once separated from the flock.

Flux is cured by the yolk of an egg boiled hard; and boiled barley soaked in wine.

Costiveness is removed by giving bran and water with a little honey; or give a small dose of castor oil.

Vermin are destroyed by giving them clean sand and ashes to roll in, adding a little quicklime if necessary.

Entire cleanliness is necessary for the avoidance of this and other diseases. A perfectly dry range is also essential, nor should there be too many together, as this is a fruitful source of disease.

THE TURKEY

This bird was unknown to the civilized world till the discovery of this Continent. It was found here both in its wild and domesticated state; and still occupies the whole range of the western hemisphere, though the wild turkey disappears as the country becomes settled. The wild is larger than the domesticated bird, sometimes weighing over 30 lbs. dressed. The color of the male is generally a greenish brown, approaching to black, and of a rich, changeable, metallic lustre. The hen is marked somewhat like the cock, but with duller hues. Domestication through successive generations dims the brilliancy of their plumage, and lessens their size and hardiness. It also produces a variety of colors, though they are mostly of a black, buff, pure white, or speckled.

They give evidence of the comparative recency of their domestication, in the instinct which frequently impels the cock to brood and take care of the young. Nothing is more common than for the male bird to supply the place of the hen, when any accident befalls her, and to bring up a family of young chicks with an equally instinctive regard for their helplessness and safety.

The flesh of this bird, both wild and tame, is exceedingly delicate and palatable; and though not possessing the high game flavor of some of the smaller wild-fowl, and especially of the aquatic, as the canvass-back duck, &c., it exceeds them in its digestibility and healthfulness. The turkey is useful principally for its flesh, as it seldom lays over a nest-full of eggs in one season, when they brood on these and bring up their young. If full-fed, and their first eggs are withdrawn from them, they frequently lay a second time.

Breeding

Those intended for breeders should be compact, vigorous, and large, without being long-legged. They should be daily, yet lightly fed through the winter, on grain and roots, and some animal food is always acceptable and beneficial to them. They are small eaters, and without caution will soon get too fat. One vigorous male will suffice for a flock of 10 or 12 hens, and a single connection is sufficient for each. They begin to lay on the approach of warm weather, laying once a day, or every other day, till they have completed their litter; which in the young or indifferently fed, may be 10 or 12, and in the older ones, sometimes reaches 20. The hen is sly in secreting her nest, but usually selects a dry, well-protected place. She is an inveterate setter, and carefully hatches most of her eggs.

The young may be allowed to remain for 24 hours without eating, then fed with hard-boiled eggs made fine, or crumbs of wheat bread. Boiled milk, curds, and buttermilk afford an excellent food. As they get stronger, oat or barley-meal is suitable, but Indian-meal, uncooked, is hurtful to them when quite young. They are very tender, and will bear neither cold nor wet, and it is of course necessary to confine the old one for the first few weeks.

When able to shift for themselves, they may wander over the fields at pleasure; and from their great fondness for insects, they will rid the meadows of innumerable grasshoppers, bugs, and beetles, which often do incalculable damage to the farmer. Early chickens are sufficiently grown to fatten the latter part of autumn or the beginning of winter, which is easily done on any of the grains or boiled roots. Both are better for being cooked. They require a higher roosting-place than hens, and are impatient of too close confinement, preferring the ridge of a barn, or a lofty tree, to the circumscribed limits of the ordinary poultry-house. When rightly managed and fed, turkeys are subject to few maladies; and even these, careful attention will soon remove.

THE PEACOCK AND GUINEA-HEN

The Peacock is undoubtedly the most showy of the feathered race. It is a native of the southern part of Asia, and is still found wild in the islands of Java and Ceylon, and some parts of the interior of Africa. They are an ornament to the farm premises, and are useful in destroying reptiles, insects, and garbage; but they are quarrelsome in the poultry-yard, and destructive in the garden. Their flesh is coarse and dark, and they are worthless as layers. The brilliant silvery green and their ever-varying colors give place to an entire white, in one of the varieties.

The Guinea-hen is a native of Africa and the southern part of Asia, where it abounds in its wild state. Most of them are beautifully and uniformly speckled; but occasionally they are white on the breast, like the Pintados of the West India Islands, and some are entirely white. They are unceasingly garrulous; and their excessively pugnacious character renders them uncomfortable inmates with the other poultry. Their flesh, though high-colored, is delicate and palatable, but, like the peacock, they are indifferent layers. Both are natives of a warm climate, and the young are tender and rather difficult to rear. Neither of these birds is a general favorite, and we omit further notice of them.

The Goose

There are many varieties of the goose. Main enumerates twenty-two, most of which are wild; and the tame are again variously subdivided. The common white and gray are the most numerous and profitable. The white Bremen is much larger, often weighing over 20 lbs. net. It is of a beautiful snowy plumage, is domestic and reared without difficulty, though not as prolific and hardy as the former. The China Goose is smaller than the gray, and one of the most beautiful of the family, possessing much of the gracefulness and general appearance of the swan. It is prolific and tolerably hardy, but has not thus far been a successful rival with the first. The Guinea or African goose is the largest of the species, and equals the size of the swan, often dressing over 25 lbs. It is a majestic and graceful bird, and very ornamental to water scenery. Several other varieties are domesticated in the United States.

Breeding

Geese pair frequently at one year old, and rear their young; but with some kinds, especially of the wild, this is deferred till two and sometimes three. They require a warm, dry place for their nests, and when undisturbed, they will sit steadily; and if the eggs have not been previously chilled or addled, they will generally hatch them all, if kept on the nest. To insure this, it is sometimes necessary to withdraw the first hatched, to prevent the old ones wandering before all are out. The young should be kept in a warm sheltered place till two or three weeks old, if the weather be cold or unsettled. The best food for the goslings, is barley or oat, or boiled Indian meal and bread. Milk is also good for them. They require green food, and are fond of lettuce, young clover, and fresh tender grass; and after a few weeks, if they have a free range on this, they will forage for themselves.

Geese are not a profitable bird to raise, unless in places where they can procure their own subsistence, or at least during the greater part of the year. This they are enabled to do, wherever there are extensive commons of unpastured lands, or where there are streams or ponds, lakes or marshes with shoal sedgy banks. In these, they will live and fatten throughout the year, if unobstructed by ice. They may be fed on all kinds of grain and edible roots, but it is more economical to give them their food cooked. The well-fattened gosling affords one of the most savory dishes for the table.

Geese live to a great age. They have been known to exceed 100 years. If allowed a free range on good food and clean water, they will seldom get diseased. When well fed, they yield nearly a pound of good feathers in a season, at three or four pluckings; and the largest varieties even exceed this quantity.

Ducks

Are more hardy and independent of attention than the goose, and they are generally the most profitable. They are omnivorous, and greedily eat everything which will afford them nourishment, though they seldom forage on the grasses like the goose, when they can procure other food. They are peculiarly carnivorous, and devour all kinds of meat, putrid or fresh; and are especially fond of fish, and such insects, worms, and other creeping things, as they can find imbedded in the mud or elsewhere. They will often distend their crop with young frogs, almost to the ordinary size of their bodies. Their indiscriminate appetite often renders them unfit for the table, unless fattened out of the reach of garbage and offensive matters. An English admiral used to resort to well-fattened rats for his fresh meat when at sea, and justified his taste by saying, they were more cleanly feeders than ducks, which were general favorites.

The most profitable for domestic use, is undoubtedly the common black duck. They lay profusely in the spring, when well fed, often producing 40 or 50 eggs, and sometimes a greater number, if kept from setting. They are much larger than those of the hen, and equally rich and nourishing, but far less delicate. They are careless in their habits, and generally drop their eggs wherever they happen to be through the night, whether in the water, the road, or farm-yard; and as might be expected from such prodigality of character, they are indifferent setters and nurses.

The ducklings are better reared by setting the eggs under a sedate, experienced hen, as the longer time necessary for hatching, requires patience in the foster-mother to develop the young chick. They should be confined for a few days, and away from the water. At first they may be fed with bread, or pudding made from boiled oat, barley, or Indian meal; and they soon acquire strength and enterprise enough to shift for themselves, if afterwards supplied with pond or river water. They are fit for the table when fully grown, and well fattened on clean grain. This is more economically accomplished by feeding it cooked.

The varieties of ducks are almost innumerable. Main describes 31, and some naturalists number over 100. Besides the black duck above described, several others, as the light gray, the white duck, and some of the tufted, are prolific, hardy, and profitable.

We omit further notice of other varieties; and of the swan, brant, pigeons, &c., as not profitable for general rearing, and only suited to ornamental grounds.

INDEX

D

E

F

G

H

I

J

L

M

N

O

P

R

S

T

U

Related Nova Publications

Zoonomia. Volume I: The Laws of Organic Life

Author: Erasmus Darwin, M.D.

Series: Animal Science, Issues and Research

Book Description: This 2 volume set is written by the grandfather of Charles Darwin. The purpose of the books is to reduce the facts belonging to animal life into classes, orders, genera, and species; and, by comparing them with each other, to unravel the theory of diseases.

Hardcover ISBN: 978-1-53614-872-5
Retail Price: $250

Zoonomia. Volume II: The Laws of Organic Life

Author: Erasmus Darwin, M.D.

Series: Animal Science, Issues and Research

Book Description: This 2 volume set is written by the grandfather of Charles Darwin. The purpose of the books is to reduce the facts belonging to animal life into classes, orders, genera, and species; and, by comparing them with each other, to unravel the theory of diseases.

Hardcover ISBN: 978-1-53614-874-9
Retail Price: $275

To see a complete list of Nova publications, please visit our website at www.novapublishers.com

V

W